The Resurgence of
Evolutionary Biology

The Resurgence of Evolutionary Biology

Ethical and Political Implications

Terry Hoy

LEXINGTON BOOKS
Lanham • Boulder • New York • Oxford

LEXINGTON BOOKS

Published in the United States of America
by Lexington Books
4720 Boston Way, Lanham, Maryland 20706

12 Hid's Copse Road
Cumnor Hill, Oxford OX2 9JJ, England

British Library Cataloguing in Publication Information Available

Library of Congress Cataloging-in-Publication Data

Hoy, Terry.
   The resurgence of evolutionary biology : ethical and political implications / Terry
Hoy.
       p. cm.
   Includes bibliographical references (p.).
   ISBN 0-7391-0263-X (cloth : alk. paper)
       1. Evolution (Biology)—Moral and ethical aspects. 2. Evolution (Biology)—Political
aspects. I. Title.

QH366.2 .H69 2001
576.8'01—dc21

                                                           2001032671

Printed in the United States of America

♾™ The paper used in this publication meets the minimum requirements of American
National Standard for Information Sciences—Permanence of Paper for Printed Library
Materials, ANSI/NISO Z39.48–1992.

Contents

Introduction/Overview

The resurgence of Darwinian theory of evolution and its implications for moral-political theory has been a striking development of the past several decades. In seeking to clarify the meaning of this development, it is necessary, at the outset, to put into broad perspective what has been the historical transition from the Greek classical era to Enlightenment and post-Enlightenment historical development in regard to defining the relation of morality to human nature. What was distinctive in this development, as George Sabine points out, was the radical break from the classical Aristotelian concept of a *telos* of nature as the unfolding of human moral ends and purpose toward a mechanistic view in which that which controls human behavior has to do with material substance obeying mechanical laws of physics. What controls human behavior is not an end, but a cause; the psychological mechanism of the human animal.[1]

The political implication of seventeenth-century mechanistic interpretation in the contribution of Hobbes and Locke provided an impetus to a concept of reason seen as instrumental to human passions and desires. According to Hobbes, a radical egoism and a war-like state is characteristic of human beings in a presocial state of nature, and it is the desire for self-preservation that leads to the development of a sovereign state as an instrument for security. Locke did not share the Hobbesian view of nature as the "war of all against all," but rather a view of nature as a source of freedom and equality; the natural rights of life, liberty, and property. The inconveniences of the state of nature, due to lack of a settled law or impartial authority, becomes the motivation for the social contract by which individuals enter into civil society for the purpose of protection of natural rights. But if Locke's theory was in opposition to

Hobbes's concept of an absolute sovereign, as Sabine observes, he shared the Hobbesian view of the fundamental egoism of human nature; a concept of pleasure and pain similar to Hobbes's concept of self-preservation.

The basic features of Lockean liberalism can thus be seen as a carryover into the nineteenth-century tradition of Benthamite utilitarianism; an ethics that is derivative from human sensations: feelings of pleasure are good, and pain is to be avoided. The interest of the community is the sum of the interests of the individuals that compose it, and the role of government is to promote the "greatest happiness of the greater number."[2] It was this optimism in the role of scientific reason as instrumental to human progress that has often been seen as the central dynamic of Enlightenment rationalism. As Ernst Cassirer noted: "the basic idea underlying all tendencies of the enlightenment was the conviction that human understanding is capable by its own power and without any recourse to supernatural assistance of comprehending the system of the world, and that new way of understanding the world will lead to a new way of mastering it."[3]

What has been a striking development of the post-Enlightenment era is a growing consciousness of the misguided optimism of Enlightenment rationalism. Few would wish to deny the genuine achievements that have resulted from the spread of constitutional government: improvements in the economic status of working classes; the extension of universal suffrage and civil liberties; technological innovations that have enhanced the quality of human life. But it would be imperative also to point to a dark side of the modernity associated with two cataclysmic wars; Western colonialism and imperialism; chronic economic crises and dislocations; totalitarian extremism of both left and right; racial and class antagonism; the now looming threat of nuclear power and the prospect of environmental destruction. By the early twentieth century, Max Weber had mounted what was to become a paradigm critique of modernity in subsequent decades. It was his view that the legacy of the Enlightenment, when unmasked and understood, was the triumph of a "purposive-instrumental rationality." This form of rationality, he believed, affects and infects the entire range of social and cultural life, encompassing economic structures, law, bureaucratic administration, and even the arts. The growth of purposive instrumental rationality does not lead to the concrete realization of universal freedom but to the creation of an "iron cage" of bureaucratic rationality from which there is no escape.[4]

A growing concern for the contradictions and ambiguities of the

Enlightenment achievement has been a strong impetus to skepticism toward the possibility of moral ideals that can be derivative from an ahistoricist, objectivist view of human nature, and the need for a reconstructive formulation that recovers the role of tradition and historical narrative in the shaping of human identity and meaning. Hans Gadamer has been an influential spokesman for how such reconstruction can be formulated. Hermeneutical interpretation, he believes, must overcome the antithesis between historicity and knowledge. What this entails, he believes, is a "hermeneutical circle" that is neither subject nor object, but rather the "interplay of the movement of tradition and the movement of interpretation." The meaning that governs our interpretation of a classical text is not an act of subjectivity, "but a consciousness of a community that binds us to a tradition." But tradition is not simply a precondition into which we come. We produce it, ourselves, insofar as we understand and participate in the evolution of a tradition. Gadamer emphasizes that this does not entail setting ourselves in the "spirit of the age" for we are always in a situation having a "horizon" or "range of vision" that includes what can be seen from a particular standpoint. The historical movement of life is not bound to any one standpoint; there are no "closed horizons." Hermeneutical interpretation is, thus, not passing into an alien world unconnected to our own. The horizon of the present, he contends, cannot be formed without the past. There is always the possibility of a "fusion of horizons."[5] While Gadamer rejects Aristotle's metaphysical biology, he perceives in Aristotle's *praxis* a model for hermeneutical interpretation. This springs from his distinction between the knowledge of *phronesis* and theoretical knowledge of *epistme*. The latter is the mode of what is mathematical, unchanging, and amenable to proof that can be known by anyone. But the knowledge of *phronesis* is the knowledge of man as acting, being concerned with what is not always the same as it is, but can also be different; "the purpose of his knowledge is to govern actions."[6]

Gadamer also points out that Aristotle is the origin of a classical view that man is a living being who has *logos*, man as *animal rationale*, distinguished from other animals by the capacity for thought. But Gadamer points out that the primary meaning of *logos* is language in which men manifest to each other what is right and wrong. It is by virtue of this fact that among man there can be a common life, a political constitution, and organic division of labor. All that is involved in the simple assertion that "man is a being who possesses language."[7]

The trend toward historical-hermeneutical interpretation has been central in the so-called communitarian critique of classical liberalism. It is the contention of Charles Taylor, for example, that a central malaise of modernity has been an "atomism" that he sees as being the product of seventeenth-century contract theory, and the primacy of rights doctrine associated with John Locke. Taylor believes it is necessary to recover a "social view of man" as one that holds that an "essential constitutive condition for a human good is bound up with being in society, community of language and mutual discourse about the good and bad, justice and injustice."[8]

Alasdair MacIntyre has become an influential spokesman for a view that the ethos of modernity, as a product of the Enlightenment, is characterized by a "moral disarray and disorder" in which disagreements about such issues as abortion, war, and justice have no point or arbitration beyond the claim of "emotivism;" in which all evaluative moral judgments are nothing but the expression of preference, attitudes, and feelings. The corrective, he believes, must be sought in the recovery of the Aristotelian concept of a *telos* as the contrast between man-as-he-happens-to-be-and-as-he-could-be if he realized his essential nature, but severed from its association with a metaphysical biology, and formulated as the understanding of moral virtues given in tradition and historical narrative.[9]

While the concept of historical-hermeneutical interpretation has gained wide acceptance in contemporary ethical-political theory, it is also subject to several serious objections. Richard Bernstein has leveled two powerful objections to MacIntyre's view that the moral bankruptcy of the Enlightenment must be supplanted by a concept of virtues given in tradition and social practices. What Bernstein finds objectionable in MacIntyre's contention is a grand Either/Or: "*Either* one must follow through in the aspirations and the collapse of the different versions of the Enlightenment project until there remains only the Nietzschean diagnosis and the Nietzschean problematic, *or* one must hold that the Enlightenment project was not only a mistake, but should never have been commenced in the first place. There is no third alternative."[10] What this entails, Bernstein contends, is an "overkill." For in Bernstein's view there is "truth in the Enlightenment prospect which itself needs to be reclaimed and preserved. We do a grave injustice to the Enlightenment if we fail to appreciate the extent to which it was a legitimate protest against hypocrisy and injustice, if we fail to appreciate how it was acutely sensitive to the failure of moral and political ideologies that systematically excluded whole groups of human beings from participating in the good life, while at the same time legitimating political beliefs that masked various forms of domination."[11]

What Bernstein finds as a further deficiency in MacIntyre's neo-Aristotelian formulation is the appeal simply to social practices as the basis for understanding of human virtues.[12] But it would be important to emphasize that this critical reaction to MacIntyre applies to the central substance of his writings up until his recent book *Dependent Rational Animals*. What is striking in this book is his confession that he was wrong in supposing that human virtues can be sustained independent of Aristotle's biology. For no account of rules and virtues can be adequate that does not take into account how forms of life are biologically constituted, starting with our animal conditions; the comparison of humans with other intelligent animal species. MacIntyre now believes this will require an effort to sustain Aristotle's view of the integral connection of human virtues with his biological theory and how this is reinforced in the contribution of Darwin. It will be the intent of this study to show how this possibility can be effectively sustained.

Such a contention is obviously controversial and problematic in view of the developments associated with nineteenth-century social Darwinism. But it will be the intent of chapter 1 to argue that the cult of nineteenth-century social Darwinism, as the emphasis upon competitive struggle and "survival of the fittest," was drawn from Darwin's account of animal evolution in his *Origin of Species* that ignores the central dynamic of his *Descent of Man*. For the keynote of this work is the human evolution from animal life giving rise to the capacities for language, sociality, and habituation, and a moral sensibility directive to a general good beyond competitive struggle, egoism, and self-interest. It was the distortions fostered by social Darwinism that gave rise to powerful critical reaction in the late nineteenth and early twentieth centuries that led to the eclipse of Darwin's influence and to the ascendance of a concept of "culturalism" promoted by leading figures in anthropology and sociology. Mid-twentieth-century developments, however, have given rise to a resurgence of the Darwinian theory of evolution. What has been central in this development has been the concept of DNA as an emphasis upon the history of life, the interplay of genetic and environmental influence. What is of central significance in this development is a framework that can be a corrective to the distortions of the Enlightenment's reductive materialism. As John Henry Randall points out: "when Darwin led men to take biology seriously, this meant the reintroduction of functional concepts that physicists had forgotten in interpreting living processes." Such a contention does not deny that all processes involve and depend on some mechanism, but that "all processes also involve the *functioning* of these mechanisms, the results they lead to, the way they cooperate with other processes."[13] Randall thus believes Darwinian theory of evolution provides a recovery of a Greek ideal: "man as a part of natural processes, but acting in ways unparalleled by any other natural being."[14]

The central core of an Aristotelian-Darwinian naturalism, it will also be argued, is a concept of human nature that can be congruent with the liberative-emancipative direction of Enlightenment and post-Enlightenment historical developments. Such a contention can be fully cognizant of how both Aristotelian and Darwinian naturalism involved prejudiced applications (Aristotle's defense of slavery and the political subordination of women; the racial and class prejudices evident in Darwin's *Descent of Man*). But such prejudiced applications, it will be contended, stood in contradiction to a concept of human nature which (from within our present-day historical horizon) we are entitled to claim as having a universality irrespective of differences of race, class, or gender that was to be given a more progressive realization in twentieth-century historical developments.

There would by no means be a general agreement that the contemporary renewal of Darwinian evolution can be defended as a continuity with Aristotelian implication. But it will be the intent of this study to draw out this implication by way of critical evolution of the main variants of contemporary Darwinian interpretation. It will be the intent of chapter 2 to outline Edward Wilson's sociobiology, which has been one of the leading contributions to contemporary Darwinian renewal: his concept of a co-genetic-cultural interaction in regard to qualities of nature's fitness such as parental investment, mating strategies, territorial expansion and defense, and social status. The ethical-political implications of Wilson's view of evolutionary biology reside in his concept of a genetic altruism as "cost-benefit" calculation that is an expectation of reciprocation. But it will then be argued that Wilson's version of a neo-Darwinian synthesis is beset by several difficulties. Wilson emphasizes the need for consilience of the Enlightenment tradition of scientific materialism with the humanities and social sciences. Yet Wilson, at the same time, concedes that higher levels of biological development cannot be reduced to lower levels, and that in regard to ethical evaluation, it is necessary to postulate human intentionality and free will. What is not clarified is that, if this is the case, the implication of evolutionary biology needs to be located more in continuity with an Aristotelian naturalism that involves an effort to avoid a mind-body dualism as well as a reductive materialism. A more serious difficulty is Wilson's conviction that it is possible to defend a cost-benefit calculation in human moral evolution for such a contention is insufficient for a concept of justice as a recognition that if society is indeed based upon reciprocity. Such reciprocity is not only one of mutual advantage but also one of mutual respect and tolerance, and an ideal of community more in congruence with Aristotelian ethical-political theory.

A third difficulty in Wilson's version of evolutionary biology is that if he is emphasizing a co-genetic cultural interaction, he insists that "genes hold culture on a leash." Such a contention thus becomes vulnerable to a charge of a no longer credible biological reductionism. It is this critique of Wilson that is central in a leading rival version of Darwinian renewal articulated by Steven Gould and R. C. Lewontin that will be considered in chapter 3. It is their central contention that the implication of Darwinian evolution must be formulated simply as a biological-cultural *interactionism*, without being able to establish what is specifically genetic versus what is cultural. What they seek to defend is a *Darwinian pluralism* in which evolutionary possibilities are simply "adaptive stories." But it will be argued that Gould and Lewontin are among the leading representatives of a Radical Left critique of Wilson's sociobiology on grounds that it perpetuates a conservative political ideology related to a past tradition of social Darwinism. What they do not clarify, then, is how their affirmation of a more pluralist Darwinism can be a basis for the defense of democratic ideas of equality and social justice beyond simply the cultural relativism of so-called adaptive stories.

It will be the intent of chapter 4 to outline Ernst Mayr's view of a Darwinian synthesis that is a mediation between Wilson's sociobiology and Gould-Lewontin's interactionism. What this entails, it will be argued, is the key to the integration of Darwinian evolution with Aristotelian naturalism. Central to Mayr's Darwinian synthesis is an emphasis upon the distinctive features of biological development that cannot be reducible to the mechanistic model of the physical sciences; his contention that what emerges at higher levels of biological development cannot be reduced to lower levels. What is of particular significance in Mayr's Darwinian synthesis is his emphasis upon the goal directive *teleonomic* features of organic life that he believes to be in continuity with Aristotle's view of the form-giving principle of developing organisms. Contemporary developments in evolutionary biology, he believes, are a confirmation of this continuity. *Teleonomic* or goal-directed behavior entails a "program," or the form-giving principle of developing organisms. *Teleonomic* or goal-directed behavior is laid down in the DNA of the genotype (a closed program) constituted to receive additional information (an open program) acquired through language and learning.

A central feature of Mayr's Darwinian synthesis mediates between Wilson versus Gould and Lewontin. This is a view that while there is a genetic basis in regard to *capacities* for ethical evaluation, the main substance of ethics (in regard to choosing alternative courses of action, the assessment of consequence) is the product of learning experiences in human growth and development. But it is Mayr's conviction that such learning experience holds out the prospect for what an exponent of evolution can anticipate as an ethics overcoming rigidities of past ethical norms, which go beyond egoism and self-interest, and directive to modern problems of overpopulation and ecological destruction.

It will be the intent of the concluding chapter to provide clarification of a framework for Darwinian-Aristotelian integration. It will be the intent, first of all, to show how a collaborative essay of Hilary Putnam and Martha Nussbaum provides an interpretation of Aristotle's naturalism within a pragmatic realism that effectively reinforces Mayr's view of the distinctive features of organic life beyond a mind-body dualism, as well as a reductive materialism. It will be contended that it is possible to sustain a convergence between Aristotle and Darwin in regard to the biological bases of ethical evolution: the relation of human to animal life, but what is distinctive to human evolution in regard to the capacities for speech, sociality, and habituation. It is Aristotle, however, who provides a more complete elaboration of the ethical-political implication only broadly indicated by Darwin himself and Mayr's interpretation of Darwinian theory. It will then be contended that the contemporary relevance of Aristotelian-Darwinian naturalism can be a corrective to both the inadequacies of the individualist-utilitarian tradition of classical liberalism, as well as what is unsatisfactory in the so-called communitarian alternatives that would invoke what is authoritative by reference to social practices and tradition. It will be contended that John Rawls's theory of justice, although commonly seen as a neo-Kantian formulation, can be effectively appropriated within an Aristotelian-Darwinian integration. For in his elaboration upon his concept of justice, Rawls speaks of an "Aristotelian principle": the view that a person's good is determined by a rational plan of life that he would choose with deliberative rationality from a maximal class of plans. Rawls also elaborates upon a concept of "natural sentiments" for justice in the process of human growth and development, and that such sentiments seem to be congruent with the evidence of human evolution.

It will be the intent finally to defend an Aristotelian-Darwinian integration in confrontation with the objection that it is an example of the so-called naturalistic fallacy that moral ideals can be deducible from non-moral biological dispositions of human nature. The response to this objection embodies an emphasis upon the concept of a "second" or "historicized nature" that is expressive of what is distinctive in human evolution in regard to *capacities* for speech, sociality, and habituation. It is such capacities that are central to the process of critical moral reflection in seeking to reconcile conflicting disposition in biological-cultural interaction (such as egoism versus cooperation) in order to arrive at consequences that can be the best approximation to a human good. But such an approximation is neither simply the subservience to biological drives and instincts nor what is above or beyond human nature. Such a contention is well put by Bernard Yak. He states that it is Aristotle's view that we certainly need law and education to live a fully human life,"But the training in virtues is not for Aristotle a fight against nature. . . . It is instead a process in which we draw out and build on human beings' natural capacities and natural impulses for communal living."[15]

Notes

1. George Sabine, *History of Political Theory* (New York: Henry Holt, 1950), 460.

2. Sabine, *History of Political Theory*, 540.

3. John Hallowell, *Main Currents in Modern Political Thought* (New York: Henry Holt, 1950), 180.

4. Richard Bernstein, *The New Constellation: The Ethical-Political Horizons of Modernity/Postmodernity* (Cambridge, Mass.: MIT Press, 1992), 36-40.

5. Hans Gadamer, *Truth and Method* (New York: Crossroad Press, 1982), 273.

6. Gadamer, *Truth and Method*, 280.

7. Hans Gadamer, *Philosophical Hermeneutics*, translated and edited by David E. Linge (Berkeley: University of California Press, 1977), 60.

8. Charles Taylor, *Philosophy and the Human Sciences, Philosophical Papers*, vol. 2 (Cambridge: Cambridge University Press, 1985), 209, 263.

9. Alasdair McIntyre, *After Virtue: A Study in Moral Theory* (Notre Dame, Ind.: University of Notre Dame Press, 1981), chapters 2, 3, 4, 11.

10. Bernstein, *The New Constellation*, 22.

11. Richard Bernstein, *Philosophical Profiles* (Philadelphia: University of Pennsylvania Press, 1986), 135.

12. Bernstein, *Philosophical Profiles*, 127.

13. Alasdair McIntyre, *Dependent Rational Animals: The Paul Carus Lectures* (Chicago: Open Court, 1999).

14. John Randall, "The Changing Import of Darwin in Philosophy," in *Darwin*, edited Philip Appleman (New York: W. W. Norton, 1979), 320.

15 Bernard Yak, *The Problems of a Political Animal: Community, Justice*

and Conflict in Aristotelian Political Thought (Berkeley: University of California Press, 1992), 15.

Chapter 1

The Rise and Eclipse of Social Darwinism

It would be obvious that if one is to understand what has given rise to the current rejuvenation of Darwinian theory of evolution, it is essential to understand how this was a continuity with Darwin's intentions, but also to clarify how this intention could have been an inspiration to nineteenth-century social Darwinism. It has been common to believe that the core meaning of social Darwinism was given classical expression by Herbert Spencer: a view of human evolution as one of competitive struggle and the survival of the fittest. But it will be the intent of this chapter to show, first of all, that Darwin's the *Descent of Man* is clearly an affirmation of what is distinctive in human moral capacities that is a break from the competitive struggle of animal life that was central in his earlier work, *The Origin of Species*. It was thus this latter work that was the main inspiration for the association of social Darwinism with the nineteenth-century ideology of laissez-faire capitalism expounded not only by Herbert Spencer but also by William Sumner and Dale Carnegie. Yet this is not to deny that *significant features* of Darwin's *Descent of Man* are indicative of how Darwin was involved in social class prejudices of the Victorian era that could serve the interests of social Darwinism. But from within our present-day historical understanding, we can effectively argue that such prejudices were not congruent with an essentially egalitarian implication in his view of the evolution of human moral capacities such as language, habituation, and natural sentiments that are expressive of a universally shared common humanity. The demise of social Darwinism in late-nineteenth- and early-twentieth-century America was influenced by social scientists who successfully exposed the distortion of scientific evidence that served to support social Darwinism

1

ideology, and who were largely exponents of cultural rather than biological determinism. But it will then be contended that the renewal of
Darwinian evolution provides a basis for cognizance of the interplay of
genetic and cultural influences that can be a basis for salvaging the
central scientific insight of Darwin's *Descent of Man*, while also sustaining a view of Darwinism as congruent with an egalitarian ideology.

I

What is central to Darwin's *Descent of Man* is his view of the features of
human evolution that are the product of natural selection in which bene-
ficial variations have been preserved and injurious ones eliminated.[1]
Man is the most dominant animal which now exists, spreading more
widely than any other organic form. His immense success, he believes, is
due to intellectual faculties and social habits that led to the development
of weapons, the art of making fire, and especially the development of
language.[2] Human bodily structure was also of great importance, including upright posture, hands, and arms that facilitated the use of tools, self-
defense, and the more complex brain structure central to the development of intellectual faculties.[3] It is these developments, then, that are
central in the emergence of human moral capacities. The emergence of a
natural moral sensibility, he contends, can be "summed up in that short
but imperious word *ought*, so full of high significance. It is the most
noble of all the attributes of man, leading him without a moments hesita-
tion to risk his life for that of a fellow creature; or after due deliberation,
impelled simply by the deep feeling of right or duty, to sacrifice it in
some great cause."[4] It was Darwin's view that any animal, once endowed
with a marked social instinct, would inevitably acquire a moral sense of
conscience, as soon as its intellectual powers were well enough developed, or nearly as well developed as in man. This contention, he believed, entails four key points. One is that social instincts lead an animal
to take pleasure in the society of its fellows, having a certain amount of
sympathy for them, and disposed to perform services for them. A second
point is that when our mental faculties have become highly developed,
this gives rise to images of past actions and motives that become a basis
for feelings of dissatisfaction in regard to results of an unsatisfied instinct. Third, after the power of language develops, individuals become
influenced by the common opinion of a community as to how an individual ought to act for the public good, and that this becomes a guide for
action. Darwin concludes, lastly, that habits will ultimately play an
important role in guiding human conduct. For social instincts are

strengthened by habits, and provide a basis for obedience to the community.[5]

While it is Darwin's contention that the social instincts of man are similar to animals, and that differences are only of degree, he recognizes that the importance of a difference lies in the fact that man may regret that he has followed one impulse rather than another. The reason for this is that man cannot avoid reflecting on past images, such as vengeance satisfied or danger avoided at cost of other men in conflict with an instinct of sympathy to his fellows that is still active in his mind. "Man thus prompted will through long habit acquire such perfect self command, that his desires and passion will at last instantly yield to his social sympathies, and there will no longer be a struggle between them." It is also possible that the habit of self-command, like other habits, can be inherited. "The imperious word *ought* seems merely to imply the consciousness of the existence of a persistent instinct, either innate are partly acquired serving him as a guide though liable to be disobeyed."[6]

Darwin is convinced that, in looking to future generations, we have no reason to fear that social instincts will grow weaker, and that virtuous habits will grow strong, fixed perhaps by inheritance.[7] Social qualities such as sympathy, fidelity, and courage, he believed, are acquired through natural selection, aided by inherited habit. But Darwin believes that another more clearly powerful stimulus to social virtues is the praise and blame of our fellow man. Primitive man, even at very remote periods, would have approved of conduct which appeared to them as contributing to a general good, and would have reprobated that which appeared evil. With increased experience and reason, man is able to perceive the more remote consequence of his action and self-regarding virtues such as temperance and charity (in earlier times disregarded) would become more highly regarded. "Ultimately a highly complex sentiment, having its first origin in social instincts, largely guided by the approbation of our fellow man, ruled by reason, self interest, and in later time, by deep religious feelings, confirmed by instruction and habit, all combined, constitute our moral sense or conscience."[8]

It would be important to emphasize that Darwin's view of human morality in his *Descent of Man* is clearly in opposition to the concept of competitive individualism that was to become the central focus of social Darwinism. In Darwin's view, human social instincts are developed rather for the general good of the community.

The term general good may be defined as the means by which the greatest possible number of individuals can be reared in full

vigour and health, with all their faculties perfect, under the conditions to which they are exposed. As the social instincts both of
man and lower animals have no doubt been developed by the
same steps, it would be advisable, if found practiceable, to use
the same definition in both cases, and to take as the test of morality, the general good or welfare of the community rather than
the general happiness.[9]

Darwin goes on to note that no doubt the welfare and happiness of the
individuals will coincide. "A contented, happy tribe will flourish better
than one that is discontented and unhappy." At an early period of history
of man, he notes, the expressed wishes of the community naturally
influenced the conduct of each member. The wish of each member for
the "greatest happiness principle" was an important "secondary guide"
while social instinct (including sympathy) serving as the "primary
guide." Darwin fully recognizes the persistence of conflict between the
judgment of the common good in opposition to particular customs and
superstitions such as "the horror felt by a Hindoo who breaks his caste,
or the shame of a Mahomeian women who shows her face." But Darwin
believes that, despite many sources of doubt, man can readily come to
distinguish between "higher and lower moral rules." The higher moral
rules are based on social instincts related to the welfare of others, supported by approbation of our fellow man and by reason; the lower relating to baser instincts having their origin in public opinion. But as man
advances in civilization and as smaller tribes unite in larger community,
"Simple reason would tell each individual that he ought to extend his
social instincts and sympathies to all members of the same nation though
personally unknown to him. This point being reached where there is only
an artificial barrier to prevent his sympathies extending to men of all
nations and races." The highest stage in moral civilization, Darwin
believes, is reached when we recognize we ought to control our
thoughts, and he quotes Marcus Aurelius: "Such as are thy habitual
thoughts such also will be the character of thy mind; for the soul is dyed
by the thoughts"[10] Darwin concludes that the social instincts acquired by
man, as well as by lower animals, are for the good of the community
based upon the wish to aid his fellow man and by some feeling of sympathy. As man became more capable of tracing the remote consequence
of his action, and where he acquires the capacity to reject "baneful
customs and superstitions," the more he becomes capable of concern for
the happiness of his fellow man. Darwin concludes, in fact, that the
social instincts (that are the principle of man's moral constitution), with

the aid of intellectual powers and the force of habit, can lead to the golden rule. "As ye who would that men should do to you, do ye to them likewise; and this lies at the foundation of morality."[11]

II

From the above survey of the ethical implication of Darwin's *Descent of Man*, it might seem implausible that his writings could be seen as a resource for defining what was to become the principle tenants of so-called social Darwinism given influential articulation by Herbert Spencer, William Graham Sumner, and Dale Carnegie. But their interpretation of Darwin's writings, it should be emphasized, was derivative largely from Darwin's *Origin of Species*, which was a primary focus upon animal rather than human evolution. A keynote of this work was a concept of the "struggle for life" by which any variation, however slight, that is profitable to an individual of any species will tend to the preservation of the individual, and will generally be inherited by its offspring. This principle can thus be called "Natural Selection." Nothing is easier, Darwin contends, than to admit to truth of "a universal struggle for life." Since natural selection acts by competition, it is not surprising that inhabitants of one country, although supposedly having been especially created and adapted in that country, are nonetheless beaten and supplanted by production from another land. "We need not marvel at the sting of a bee causing the bee's own death; at drones being produced in such vast numbers for one single act, and then being slaughtered by their sterile sisters . . . the instinctive hatred of the queen bee for her own fertile daughters. . . . The wonder indeed is, on the theory of natural selection, that more cases of the want of absolute perfection have not been observed."[12] Darwin believed that the process of natural selection is also apparent in sexual selection. While this does not depend on the struggle for existence, it does involve the struggle between males for possession of the female, which, if not leading to death of the unsuccessful competitor, does mean few or no offspring. Sexual selection is therefore more rigorous than natural selection. The most rigorous males, those best fitted for their place in nature, will leave the most progeny.

What Darwin saw to be inherent in the process of natural selection is that it was for the good of each being, and that all corporal and mental endowment will lead to progress toward perfection.

Thus from the war of nature, from famine and death, their most exalted object which we are capable of conceiving, namely the

production of the higher animals, directly follows. There is gran-
deur in this view of life, with its several powers having been
originally breathed into a few forms or into one; and that whilst
this planet has gone cycling on according to the fixed law of
gravity, from so simple a beginning endless forms most beautiful
and most wonderful have been and are being evolved.[13]

It is not difficult to see how the main substance of social Darwinism, as
an ideological support for post-Civil War American economic development, could be inspired by the concept of Darwin's natural selection as
outlined in his *Origin of Species*. As Richard Hofstadter points out, "The
survival of the fittest was a biological generalization of the cruel process
which reflective observers saw at work in early nineteenth-century
society, and Darwinism was a derivative of political economy."[14] It was
Herbert Spencer, Hofstadter believes, who was a leading figure in the
application of Darwinian theory. "His categorical repudiation of state
interference with the 'natural,' unimpeded growth of society led him to
oppose all state aid to the poor. They were unfit, he said, and should be
eliminated." Spencer deplored not only poor laws, Hofstadter points out,
but also state-supported education, sanitary supervision other than the
suppression of nuisance, regulation of housing conditions, and even state
protection of the ignorant from medical quacks. He also opposed tariffs,
state banking, business enterprises, and government postal systems.[15]

Hofstadter believes William Graham Sumner was the most vigorous
and influential social Darwinist in America. The progress of civilization,
Sumner believed, depends upon the selection process that in turn depends upon the process of unrestricted competition. "Competition is the
law of nature which can no more be done away with than gravitation and
which man can ignore only to their sorrow." Principles of social evolution, Sumner believed, negated the American ideology of equality and
natural rights. For "there can be no rights against nature except to get out
of her what we can which is only the fact of the struggle for existence
stated over again. All claims of rights, far from being absolute, are
antecedent to a particular culture, is an illusion of philosophers, reform-
ers, or anarchists."[16]

Hofstadter points out that successful business entrepreneurs "ac-
cepted almost by instinct the Darwinian terminology which seemed to
portray the conditions of their existence." The most prominent of the
disciples of Spencer was Dale Carnegie. Just as Spencer, Carnegie
emphasized the biological foundation of the law of competition. For it
cannot be evaded: "No substitutions for it have been found; and while

the law may sometimes be hard for the individual, it is best for the race because it insures the survival of the fittest in every deportment." All progress, Carnegie believed, has been displacement of communism by an intense individualism; the accumulation of wealth by those who possess the ability to produce it.

> We might as well urge the destruction of the highest existing type of man because he failed to reach our ideal as to favor the distinction of individualism, private property, the Law of Accumulation of Wealth, and the Law of Competition; for these are the highest results of human experience; the soil in which society so far has produced the best fruit. Unequal or unjustly, perhaps, as these laws sometimes operate, and imperfect as they appear to the Idealist, they are, nevertheless, like the highest type of man. The best and the most valuable of all that humanity has yet accomplished.[17]

If the biological reductionism of Darwin's *Origin of Species* could be of inspiration to the ideology of laissez-faire capitalism, it could also be supportive to further features of social Darwinism associated with theories of antisocial behavior, criminality, and feeblemindedness. This development has been well documented by Carl Degler. An influential example was Henry Goddard's study of the Kallikok family, which constituted a case study in the hereditary basis of feeblemindedness.[18] Degler also points to the emergence of a concept of "eugenics" which led a committee of the American Breeders Association to investigate a report on heredity in the human race. On the eve of World War I, eugenics became a fashionable reform movement on both sides of the Atlantic. Degler refers to an article in the professional *Journal of Eugenics* that calls on us to bear in mind that the theory of "acquired characteristic" is no longer tenable and that we should remember Darwin's admonition that an "improved environment tends ultimately to degrade the race by causing an increase survival of the unfit."[19] Degler also points to a reform movement committed to legal authorization of sterilization of the so-called unfit. Of even wider significance was the movement in the 1920s to use mental tests as a basis for ethical-racial differences, which became the basis for restrictions placed on immigration in the Immigration Act of 1924.[20]

III

If, as indicated above, exponents of social Darwinism could find support for their contention in Darwin's *Origin of Species*, then they were doing so as an interpretation of what he saw to be features of animal life that neglects what he saw to be the distinctive qualities of human evolution in his *Descent of Man*. Yet this is not to deny that there were *features* in *Descent of Man* that could be supportive of particular components of social Darwinist ideology. This is particularly evident in Darwin's discussion of differences between "civilized" and "non-civilized" races. Darwin comments that among savages, the weak mind is soon eliminated, while "we civilized men build asylums for the imbecile, the maimed, and the sick, and institute poor laws and medical care. This development is the incidental result of the instinct of sympathy that has become more widely diffused."[21] This feature of Darwin's view of human moral development is not intrinsically a manifestation of class or racist prejudices, but it clearly takes on the implication in his sympathy for views of "Mr. Greg Galton" that the obstacle to members of a superior class is due to the presence of the "poor and reckless, degraded by vice, marrying early and producing more children; the careless, squalid, unaspiring Irishmen that multiply like rabbits."[22]

Where Darwin's racist bias becomes most apparent is in his comment on the "remarkable success of the English as colonist over other European nations, their "daring and persistent energy . . . . There is apparently much truth in the belief that the wonderful progress of the United States, as well as the character of the people, are the result of Natural Selection. The more energetic Western and courageous men from all parts of Europe having emigrated during the last ten or twelve generations and having there succeeded best."[23] It is not difficult to see how this component of Darwin's *Descent of Man* could be an inspiration to the racist component of social Darwinism that was rampant in the early part of the nineteenth century among American social scientists. As Hofstadter points out: "Darwinism was put in the service of the imperial urge. . . . The measure of world domination already achieved by the race seems to prove it the fittest."[24] It was Theodore Roosevelt, he points out, who held the view that the "frontiersmen struggle with the Indian was not to be stayed and a racial war to the finish was inevitable." John Fiske, he also notes, was one of the earlier "synthesizers of evolutionism, expansionism, and the Anglo Saxon myth."[25] Fiske long believed Aryan racism to be superior to the "Teutonic theory of democracy." Fiske believed the great population potential of the English and American races was evident in a manifest destiny, "where man would finally pass out of barbarism, and become truly Christian."[26] Reverend Josiah Strong, Hofstadter comments, was an even stronger proponent of the superiority of the Anglo-Saxon race as embodying the consummation of human progress. Strong believed that a new and finer physical type was emerging in America that Darwin himself had employed in his *Descent of Man* (as noted above). When unoccupied lands are filled up, Strong contended, the world will enter a new stage of history: "the final competition of races for which the Anglo-Saxon is being schooled."[27] Hofstadter believed the fight for annexation of the Philippines was a significant expression of American Manifest Destiny celebrated by political leaders such as Albert Beveridge, Henry Cabot Lodge, John Hay, and Theodore Roosevelt. Speaking before the Senate in 1899, Beveridge commended that "God had been preparing the English speaking and Teutonic nations as master organizers of the world to overcome chaos so that we may administer government among savages and senile peoples."[28] Theodore Roosevelt emphasized the responsibilities of the United States in Hawaii, Cuba, Puerto Rico, and the Philippines. This, he said, calls for "strenuous endeavor" in order to insure that the "bolder and stronger people will pass us by and will will for themselves the domination of the world." The impulse to expansion, John Hay contended will triumph: "No man, no party can fight with any chances of final success against a cosmic tendency; no cleverness no popularity avails against the spirit of the age."[29]

It is not surprising that the ideology of social Darwinism could have such a powerful influence in the context of early-nineteenth-century American economic development. But Hofstadter points out that it would also be hard to exaggerate the temper of change in the later part of the century due to the humanitarianism of the day, the political reassessment of the common man, and the new sociology that was stimulated by the rising current of Progressivism.[30] What accompanied this development, as Carl Degler emphasizes, was a critical reaction to social Darwinism by psychologists, sociologists, and anthropologists, who, if concerned with the scientific deficiencies of social Darwinism, were also strongly motivated by an ideological aversion to violations of rights of ethnic and racial minorities that had been perpetrated by social Darwinism.[31]

Degler points to Franz Boas as a key figure in this development. "Boas' influence upon American social scientists in matters of race can hardly be exaggerated. At the same time that racial segregation was being imposed by law in the states of the American South, and eugenics was emerging as a hereditarian solution to social problems, Boas was embarking on a life-long assault on the idea that race was a primary source of the differences to be found in the mental or social capabilities of human groups. He accomplished his mission largely through his ceaseless, almost relentless articulation of the concept of culture."[32] Boas, Degler points out, denied the existence of innate differences between savage, colored, and civilized white people. Difference in physical appearance lead to no significant difference in mental or social function. He challenged the view that head shape, for example, was a basis for different racial groups in America. In his view, the shape of the head undergoes far-reaching changes after the transfer from European to American soil, which indicated that social environment must have been the source of change. Differences between civilizations of the old world, he believed, are simply the result of differences in time and do not justify any assumption that races which developed more slowly were less gifted. Boas challenged Spencer's view that primitive people were "impulsive" or lacked self-control seen as "indispensable" to civilized people. Such a contention is little more than the optimism of someone who also believed in "stability of the existing conditions."[33] From his own experience among Eskimos, Boas also challenged the traditional view that primitive peoples do not enjoy beauty as we do, for he "expresses his grief in mournful song and appreciates humorous conception." The mind of the savage is, in fact, sensitive to the beauties of poetry and music, and that "it is only the superficial observer to whom he appears stupid and unfeeling."[34] Boas also constantly challenged the view about inferiority of African Americans based upon assumptions of small brain size. He said nothing is known about differences in structure that might result from differences in size. He believed such a relation to ability is only an analogy, and that we find a number of great men with slight brain weight.[35] Boas also challenged those who contended that blacks were incapable of fitting into the complex order of the United States. For anthropologists, in his view, show that traits of African culture as observed in the aboriginal home of blacks show a considerable degree of industry, talent for organization, imagination, technical skill, and thrift.[36]

Degler believes that Alfred Kroeber picked up where Boas left off in making culture the alternative to racial explanation and independent from biology. Kroeber thus rejected the known view that natural selection affected culture for "civilization introduces a factor "practically or entirely lacking in existence of animals or plants, and consequently untouched by natural selection." Human civilization has changed profoundly without accompanying natural changes in the human organism and whereas animals adjust to environmental change through physical alteration, human beings do so through culture and history.[37]

Degler also notes the increasing number of scholars in psychology, sociology, and education in the later part of the century who challenged the traditional Darwinian view of sexual difference, and its relation to natural selection. Among some critics, this involved the contention that females in certain respects are actually biologically superior to males as a primary unit of creation, such as the bonding essential to human society; the limited role of males in continuation of the species; the qualities peculiar to women in regard to cooperation and nonaggressiveness; the evidence of higher achievement of women during the period of their greater access to higher education. But the main thrust of the critical reaction to sexism was simply the emphasis upon cultural as opposed to biological influences in shaping behavior and the process of social conditioning and that lack of opportunity accounted for differences.[38] Degler notes the importance of the contribution of Ruth Benedict and Margaret Mead in the 1930s as a challenge to sexist ideology. In her study of three primitive societies Benedict concluded that all societies have "institutionalized the roles of men and women" but that this shows wide variability. Among the Arapesh, for example, men as well as women were cooperative, unaggressive, responsive to the needs of others. But both men and women among the Mundugumor were aggressive and ruthless, having a stronger interest in sex. Among the Tchambuli, women were "dominant, impersonal, and bossy," while men were emotionally dependent. The conclusion is thus inescapable that "such differences are due to social conditions and how children are raised."[39]

Margaret Mead's studies, Degler points out, also contributed to the view that human nature is "almost unbelievably malleable." In her *Male and Female*, this entailed the contention that even behavior patterns related to biological reproduction had been shaped by culture. Her study of the Arapesh and Mundugumor, she believed, indicates that "learned behavior had replaced biologically given ones."[40]

The general trend of this period, Degler points out, was a radical "decoupling of human behavior from instinct, and the view that all responses of both animals and humans are in some fashion learned, rather than inherited."[41] By the 1930s, Degler contends, discussion of instinct had almost disappeared from psychological journals and the realization that any tests identifying differences between races or ethnic groups could no longer be considered credible. Even a leading exponent of the U.S. Army tests, Carl Bregham, now conceded that in comparing groups, he had learned that "tests is the vernacular must be used only with individuals having equal opportunities to acquire the vernacular of the test."[42] A strong challenge was also directed against the view of difference in intelligence between American blacks and whites. What was of particular significance in this challenge was a study by Otto Klineberg of black children in New York City, the result of which demonstrated that the longer children had lived there, the higher their scores. "Environment, alone, not heredity, or selective migration accounted for these gains."[43] Klineberg further contended that the real test of black-white equality in regard to intelligence tests can be answered only by the study of a region in which blacks suffer from discrimination. Klineberg believed this was approximated in Martinique or Brazil. He then cited a study of black Jamaica in which the gap between the mental ratings of whites was narrower than that resulting from comparisons in the United States. He concluded that it is safe to say that as the environment of blacks approximate more closely that of whites their inferiority tends to disappear.[44]

In commenting upon the final triumph of civilization against biological explanations during the first part of the twentieth century, Degler wonders why this development was so decisive. Certainly criteria of scientific and professional investigation were crucial, but Degler believes it is necessary to look beyond this fact to a powerful demand for equality of opportunity, and the realization that, if biological inferiority were to be proved, then groups designated as biologically inferior would be denied equal rights.[45] Degler points out, for example, that Boas did not arrive at his position from a "disinterested scientific inquiry," but from an ideological commitment in his early life. This does not mean that this caused him to distort evidence against racist doctrine. But there is no doubt he was deeply interested in collecting evidence and providing arguments to refute the ideology of racism.[46] Degler gives a further example of the view of anthropologist Alexander Goldenweiser's acknowledgement of his ideological commitment in admitting that in 1925 black Africa may have not achieved the level of science and the quality of religion of Europeans. But this does not call into question the unity of human nature. "Will we ever accept . . . the Mongol, the Indian, the Arab, the Negro . . . as equals? Who knows . . . who can doubt that we *should*."[47]

Nazi racism evident in the Holocaust, Degler believes, gave impetus to the attack upon eugenics. A significant indication of this development, for example, was the unanimous support for the 1938 resolution of the American Anthropological Association that there is no scientific basis for discrimination against people on the basis of racial inferiority or religious affiliation or linguistic heritage.[48] The UNESCO statement in 1950, Degler also notes, included a statement that biological studies support an ethics of universal brotherhood.[49] The period of the 1930s also showed significant decline in articles in professional journals in discussing heredity of racial differences. Degler comments: "No work spread the word of the triumph of culture more broadly or effectively than Ruth Benedict's *Patterns of Culture*. In her view, 'there has never been a time when civilization stood more in need of individuals who are culture-conscious; who can see objectively the socially conditioned behavior of other people without fear and recrimination.'" In her mind, Degler contends, a "clear recognition of the cultural basis of race prejudice is a desperate need in present Western Civilization.'[50]

IV

It may appear surprising that in light of the demise of social Darwinism and the triumph of culturism in the early part of the twentieth century that there could be a significant renewal of Darwinian evolution in the period since the 1960s. Phillip Appleman well sums up several examples of what has contributed to this renewal. One is the discovery of hominoid fossils in East Africa that has pushed back proto-human evolutionary history millions of years.[51] A second even more significant development has been DNA research in the 1970s that has provided a new insight into the evolutionary history of life.[52] A third influence is the ethology of the interplay of genetic and environmental influences that also rests on a theory of evolution popularized in such books as *The Naked Ape*, and *The Territorial Imperative*.[53] A fourth influence has been the new discipline of sociobiology given most influential articulation by E. O. Wilson: the study of the biological bases of every kind of organism, which has often been referred to as the "modern synthesis."[54]

It is important to emphasize that the current trend toward revival of biological explanation represents a strong emphasis upon ordinary reductionism and rather the concern for an *interdependence* of biological and cultural explanations. As Degler points out, this is well illustrated in Alfred Somit's view of the implication of ethology for political understanding: the emphasis that "*some* political phenomena are due to *some* measure of genetic programming; that an acknowledgement of role of biology in human action does not commit us to denying other influences."[55] Roger Masters notes the same point: an ethological theory that is a bridge to biology, that is *not* determined, not reductionist, a biological approach to human behavior that can be presented in a probalistic and nondeterminist form.[56] The contemporary trend in renewal of evolutionary biology, as Degler points out, is not an attempt to renew repudiated ideas or racism, sexism, and eugenics, or an attempt to order society according to hierarchies of other normative outlooks. The return of biology does not "resuscitate Herbert Spencer. Rather, the true aim of those social scientists who advocate a 'return' is to place once again, the study of human nature within evolution, to ask how human beings fit into that framework which Darwin laid down over a century ago and which very few social scientists consciously repudiated."[57]

But it would be important, finally, to respond to the objection that there remains an incoherence in the contemporary renewal of Darwinism's evolution that is at some time a rejection of social Darwinism. It was, as noted above, of the social Darwinism of Spencer as derivative of Darwin's *Origin of Species* as a focus upon animal behavior that ignores what Darwin's *Descent of Man* entails as distinctive features of human evolution in which ethics evolution constitutes a break from the cosmic process of animal life. Yet it was also noted that features of *Descent of Man* reveal racist and class implications that could be of service to exponents of social Darwinism. But there the essential point to be made is that such features give evidence of his inability to escape the Victorian prejudices of his time that simply do not follow from his scientific view of human evolution that are clearly a view of qualities of a common humanity irrespective of class, sex, or racial difference. Faulting Darwin on this point is not to contend scientific inquiry is a neutral, disinterested investigation of empirical phenomena. It was noted above that scientific protest against social Darwinism in the late nineteenth and early twentieth centuries is a clear indication of how scientists were strongly motivated by moral revulsion against threats to democratic ideals of equality in American life. This involves the conviction, as Michael Ruse points out, that the claims of scientists are expressive of the way they feel the world *should* be and *must* be. Yet this does not mean science can be a warrant for the view that "anything goes;" their claims must stand or fall by general criteria of "good science" and confirmation of facts. Where we can fault Darwin, then, is that the ideological components of his *Descent of Man* were not consistent with his scientific view of the evolution of human moral capacities which clearly entails an essentially egalitarian view of the capacities common to a universal humanity. This implication, it can thus be argued, has been effectively clarified in the current Darwinian renewal. This is not to deny that Darwinism has an "ideological" component. But as Michael Ruse contends, it is an ideology that we are all of equal worth as human beings. The Darwinian model applies to all people--black and white, preliterate, rich and poor.

All human beings are joined, being the product and still subject to the same causal models. The Kalahari bushman and the New York business executive are not alien beings. For all their differences they are brothers under the skin, united by their evolutionary predicament. They love, hate, play, work, help, fear, worship for the same reasons. They are members of the same species: Homo sapiens.[58]

Notes

1. Charles Darwin, *The Descent of Man and Selection in Relation to Sex*, with an introduction by John Tyler and Robert M. May (Princeton, N.J.: Princeton University Press, 1981), 136.

2. Darwin, *Descent of Man*, 137.

3. Darwin, *Descent of Man*, 138.

4. Darwin, *Descent of Man*, 70.

5. Darwin, *Descent of Man*, 73.

6. Darwin, *Descent of Man*, 91.

7. Darwin, *Descent of Man*, 104.

8. Darwin, *Descent of Man*, 166.

9. Darwin, *Descent of Man*, 98.

10. Darwin, *Descent of Man*, 101.

11. Darwin, *Descent of Man*, 106.

12. Charles Darwin, *Origin of Species*, edited with an Introduction by J.W. Burrow (New York: Penguin Books, 1968), 445.

13. Darwin, *Origin of Species*, 460.

14. Richard Hofstadter, *Social Darwinism in American Thought* (New York: George Braziller, 1955), 38.

15. Hofstadter, *Social Darwinism*, 41.

16. Hofstadter, *Social Darwinism*, 59.

17. Andrew Carnegie, *The Gospel of Wealth in Darwin*, edited by Philip Appleman (New York: W. W. Norton, 1979), 400.

18. Carl Degler, *In Search of Human Nature: The Decline and Revival of Darwinism in American Social Thought* (New York: Oxford University Press, 1991), 37.

19. Degler, *In Search of Human Nature*, 43.

20. Degler, *In Search of Human Nature*, 52.

21. Darwin, *Descent of Man*, 168.

22. Darwin, *Descent of Man*, 174.

23. Darwin, *Descent of Man*, 179.

24. Hofstadter, *Social Darwinism*, 173.

25. Hofstadter, *Social Darwinism*, 176.

26. Hofstadter, *Social Darwinism*, 177.

27. Hofstadter, *Social Darwinism*, 179.

28. Hofstadter, *Social Darwinism*, 180.

29. Hofstadter, *Social Darwinism*, 181.

30. Hofstadter, *Social Darwinism*, 202-204.

31. Degler, *In Search of Human Nature*, 188.

32. Degler, *In Search of Human Nature*, 61.

33. Degler, *In Search of Human Nature*, 69.

34. Degler, *In Search of Human Nature*, 70.

35. Degler, *In Search of Human Nature*, 76.

36. Degler, *In Search of Human Nature*, 77.

37. Degler, *In Search of Human Nature*, 95.

38. Degler, *In Search of Human Nature*, 117.

39. Degler, *In Search of Human Nature*, 134.

40. Degler, *In Search of Human Nature*, 134.

41. Degler, *In Search of Human Nature*, 159.

42. Degler, *In Search of Human Nature*, 176.

43. Degler, *In Search of Human Nature*, 181.

44. Degler, *In Search of Human Nature*, 183.

45. Degler, *In Search of Human Nature*, 188.

46. Degler, *In Search of Human Nature*, 189.

47. Degler, *In Search of Human Nature*, 203.

48. Degler, *In Search of Human Nature*, 204.

49. Degler, *In Search of Human Nature*, 206.

50. Philip Appleman, "Darwin: On Changing My Mind," in *Gospel of Wealth*, 551.

51. Appleman, "Darwin," 552.

52. Appleman, "Darwin," 239.

53. Appleman, "Darwin," 239.

54. Degler, *In Search of Human Nature*.

55. Degler, *In Search of Human Nature*, ix.

56. Degler, *In Search of Human Nature*, ix.

57. Degler, *In Search of Human Nature*, ix.

58. Michael Ruse, *Darwin Defended: A Guide to the Evolution of Controversies* (London: Addison-Wesley Publishing Co., 1982), 280.

Chapter 2

Wilson: Sociobiology

The sociobiology of Edward Wilson appears in the context of developments since the 1940s that, as noted in the introduction, have led to a significant renewal of evolutionary biology. What this entails, it has been seen, is the emergence of social scientists who were beginning to have serious doubts about triumphs of culturalism in the early part of the century, and who were disposed to take a "second look" at more biological-hereditary explanations of human behavior. This development, it was noted, was influenced by DNA research, and developments in ethology as the emphasis upon the interplay of genetic and environmental influences in the understanding of human evolution. Wilson defends a new discipline of sociobiology as the study of the biological basis of social behavior in every kind of organism, including man, which is being pieced together with contributions from biology, psychology, and anthropology. There is, of course, nothing new about analyzing social behavior and even the word, sociobiology, has been around for some years. What is new, he believes, is the way "facts and ideas are being extracted from their traditional matrix in psychology and ethology, and reassembled in compliance with the principles of genetics and ecology."[1]

Although Wilson has, generally, been recognized as one of the leading figures in the contemporary renewal of Darwinian evolution, his contribution has been highly controversial and seen by some critics as providing support for a conservative political agenda, having a continuity with the past tradition of social Darwinism. It will be the intent of this chapter to stress, first of all, Wilson's general view of Enlightenment science as a formulation for unified learning essential to the understanding of human evolution, and how this possibility has been eroded

by the fragmentation of knowledge, the trend toward excessive specialization, and the influence of postmodern skepticism. It will be the intent,
second, to outline his view of developments in neuroscience, behavioral
genetics, and evolutionary biology that, he believes, provide a hopeful
prospect for the "consilience" of natural science with the social sciences
and the humanities. It will be the intent, third, to outline Wilson's view
of the concept of "Natural Fitness," and to show that what this entails is
not a conservative apologetic, but rather an indication of dispositions
that we must "work around" in achieving public policy objectives. It will
be the intent, fourth, to show how Wilson seeks to establish the ethical
implications of genetic altruism, and what he believes is directive to a
conservation ethics necessary to confronting contemporary problems of
poverty and environmental destruction. But it will then be argued that if
Wilson's version of Darwinian renewal cannot be fairly accused of being
in service to a conservative ideology, it is beset by several difficulties.
One is simply what is contradictory in his celebration of the tradition of
scientific reductionism, while yet acknowledging that higher levels of
organic life cannot simply be reducible to lower levels, and his affirmation of the need to sustain a concept of free will and human intentionality in moral evaluation. It will be the intent, also, to emphasize the
inadequacy of Wilson's view of a genetic basis for a "cost-benefit reciprocal altruism" as a basis for defining what is essential to both personal
relationships and political justice. It will be argued finally that, although
Wilson is fully committed to a concept of co-genetic-cultural interaction
in human evolution, his insistence that "genes hold culture on a leash"
makes him vulnerable to the charge that he remains committed to an
untenable biological reductionism.

I

In considering Wilson's sociobiology, it is essential to show how it
grows out of his larger view of the contribution of the heritage of
Enlightenment science, as the possibility for the unification of knowledge; the developments that have eroded this possibility; but the developments that are also hopeful in the consilience of Enlightenment
science with the social sciences and the humanities. The early Enlightenment, Wilson believes, was a "giant dream of scientific unity" given
most notable articulation by Condorcet's vision of human progress. Yet
Wilson is fully cognizant of the tragic historical events that discredited
the Enlightenment's faith in scientific reason. "Might its idealism have
contributed to the terror which foreshadowed the horrendous dream of

the totalitarian state. If knowledge can be consolidated, so might the
'perfect' society be designed—one culture, one science—whether fas-
cist, communist, or theocratic."[2] Wilson contends that the Enlightenment
was not a unified movement, and what was distinctive in its achievement
was the agreement upon the "power of science to reveal an orderly and
understandable universal, and thereby lay an enduring basis for free
rational discourse." At the forefront of the Enlightenment, Wilson believes, was a group of scientists that included Bacon, Hobbes, Hume,
Locke, and Newton. Science was the "engine of the Enlightenment." It
was Bacon who challenged medieval scholasticism, affirming the principle of inductive reason, foreshadowing modern developments in the
social sciences and parts of the humanities. While he believed repeated
testing of knowledge by experiments, this meant more than manipulation
in the manner of modern science. His method of induction rejected
divisions among disciplines, a method that starts from description of
phenomena, but proceeds to higher levels of perception of the world that
included the psychological realm and the play of emotion; the use of
aphorism, fables, and analogies. In his concern for reform and reason
across all branches of learning, Bacon wished to perform reasoning
against idols of the mind; the tribe, the marketplace, and the theater:
"Stay clear of these idols, he urged, observe the world around you as it
truly is, and reflect on the best means of transmitting reality as you have
experienced it; put into it every fiber of your being."[3]

Wilson believes that Newton was a further leader of the Enlightenment "who answered Bacon's call." He did so through his invention of
calculus, his view of the external laws of method, discovered by manipulation of physical processes, his formulation of mass and distinct laws of
gravity, and his exposition of the three laws of motion. Western society
thus took the lead in worldwide development because it cultivated
reductionism and physical laws to "expand the understanding of space
and time beyond that attainable by the unaided senses."[4]

But, Wilson believes, this development had a negative side in which
the realism of the universe seemed to "grow progressively more alien."
Such a development has been the conquest of twentieth-century relativ-
ity and quantum mechanics.

The physicists succeeded magnificently, but, in so doing, they revealed the limitation of intuition unaided by mathematics; an understanding of nature, they discovered, comes very hard. Theoretical
physics and molecular biology are acquired tastes. The cost of scientific
advance is the humbling recognition that reality was not constructed to
be easily grasped by the human mind. This is the cardinal tenet of scien-

tific understanding: Our species and its way of thinking are a product of evolution, not the purpose of evolution.[5]

Wilson notes that while the radical Enlightenment was an attack upon classical theism, it provided an alternative formulation in the concept of Deism: a view of God that is material, rather than personal. But traditional theism in its emphasis upon both reason and revelation remained more appealing. "The fatal flaw in deism is thus not rational at all, but emotional. Pure reason is unappealing because it is "bloodless," lacking a sense of the "importance of ceremony that loses its emotional force where stripped of social mystery. There can thus be no substitute for surrender to an infallible and benevolent being, the commitment called salvation; the leap of faith called transcendence. "It follows that most people would very much like science to prove the existence of God but not to take the measure of his capacity." Wilson comments that Deism of science also failed to "colonize ethics," and that the Enlightenment's promise of an objective basis for formal reasoning could not be met.[6]

Wilson believes there was another more purely rationalist objection to the Enlightenment program. While the Enlightenment entailed a "Promethean view that knowledge can liberate mankind by lifting it from above the savage world," there was also the consciousness of a more "Faustian" implication: the possibility of humanity losing its freedom through totalitarian ideologies such as what occurred in the French Revolution, and the more recent theory of scientific socialism and racialist fascism. Fear of the Faustian side of science gave rise to the Romantic revolution that celebrated spontaneity, intense emotions, and heroic visions; a development inspired by writers such as Rousseau, Goethe, Hegel, and Schelling. The result was that natural scientists, intimidated by the objections to their agenda, abandoned the examination of human life, giving free play to philosophers and poets. By the early 1800s, Wilson contends, the "splendid image of the Enlightenment" was fading. For what ensued was a trend away from unification. It was, in fact, the success of reductionism that was partly the reason for this decline. For because of the accelerating growth of scientific information, research became less concerned with unification, even less with philosophy. Another reason was simply the lack of interest and commitment by scientists who saw no profit in discoveries that have a broader focus. "It was thus not surprising to find physicists who do not know what a gene is, and biologists who guess that string theory has something to do with violence. Grants and honors are given to science for discoveries, not for scholarship and wisdom."[7] The same professional atomization, Wilson believes, afflicts the social sciences and humanities. "The faculties of higher education around the world are a congress of experts. To be an original scholar is to be a highly specialized world authority in a "polyglot Calcutta of similar focused world authorities. . . . Professional scholars in general have little choice but to 'dice up' research expertise and research agenda among themselves. To be a successful scholar means spending a career in membrane biophysics, the Romantic poets, early American history, or some other such restricted area of formal study."[8]

The break from unification in the natural sciences, Wilson believes, is also accompanied by the disjunction of the social sciences from the natural sciences. What people hope from the social sciences such as anthropology, sociology, economics, and political science, is the knowledge people expect that can be a basis for lives and controlling their future.[9] But this cannot be possible without embracing the natural sciences. The current status of the social sciences he contends can be best clarified by comparison with medical science. Both have been confronted by urgent problems, but if progress in medical science has been dramatic, the same is not true of the social sciences. The failure of social sciences, he believes, is due to specialization, disunity, and a failure of vision. Progress in the social sciences has been hampered by a historical process that sought to isolate these disciplines from biology and psychology: a development that was distorted by political ideologies because of their association with eugenics, racism, and social Darwinism. It was this reaction that influenced a trend toward a cultural relativism. During the 1960s and 1970s this development was reinforced by "multiculturalism" or "identity politics" in which ethnics, women, and homosexuals have subcultures deserving of equal study. Such a contention, Wilson acknowledges, has a "good measure of reasonableness" for what can be wrong with identity politics if it increases civil rights? Thus, "many social scientists, fortified by their humanitarian concerns, grew stronger in support of cultural relativism, and they stiffened their opposition to biology in any disguise."

Such a contention, Wilson believes, leads to the conclusion: "So no biology." It is a conclusion that turns against the ideal of a unified nature grounded in heredity, and leads to an ideologically dangerous implication. For what, then, can be the basis for a unified humanity? This question cannot be 'left hanging.' For if ethical standards are simply the product of endless cultural diversities, what disqualifies theocracy, for example, or colonialism, child labor, torture, or slavery?[10]

Wilson comments that all movements tend toward extremes which is where we are today with the opposition of the Enlightenment to postmodernism. The difference between the two extremes can be expressed roughly as follows:

> Enlightenment thinkers believe we can know everything and radical postmodernists believe we can know nothing. Reality is a construction of the mind: there are no objective truths but only versions, articulated by ruling social groups. One culture is thus as good as any other in the expression of truth and reality; political multiculturalism is thus justified. In the area of literary criticism, Derrida has been a leading protagonist of deconstruction: there is nothing outside the text.[11]

In his discussion of developments in the social sciences, Wilson notes the ascendency of hermeneutics or the close analysis and interpretation of texts where each topic is examined by many scholars of differing viewpoints and cultures. Wilson acknowledges this development marks the best of "natural history" through large sectors of biology, geology, and other branches of the natural sciences. But what it lacks is an effort to explain phenomena by "webs of causation across adjacent levels of organization." This requires the aligning of their explanations with the natural sciences. What this requires must avoid Richard Rorty's conclusion that discourse can proceed "without worrying about consilience." In Wilson's view: "Creativity in research can occur unexpectedly in any form of inquiry, of course, but to resist linking discoveries by causal explanation is to diminish their credibility. It waves aside the synthetic scientific method, demonstrably the most powerful instrument hitherto created by the human mind. Lazily, it devalues intellect."[12]

II

Wilson is convinced that, despite the developments that have eroded the early Enlightenment dream of unity, there are recent developments that hold out a promise for the conciliation of natural science with the social sciences and the humanities. Physics, chemistry, and biology, he believes, have been connected by causal explanation. Quantum theory, for example, underlies atomic physics, "which is the foundation of regent chemistry and its offshoot biochemistry which interacts with molecular biology." A sequence of explanations thus proceeds from general phenomena to more complex and specific phenomena. "Such is the unifying and high productive understanding of the world that has evolved in its success testified to a fortunate combination of three circumstances: The surprising orderliness of the universe, the possible intrinsic consilience of all knowledge concerning it, and the ingenuity of the human mind in comprehending both."[13] What this development entails, he believes, can now extend to the social sciences and the humanities, calling the traditional division of knowledge into question. At the heart of this possibility is the relation of culture to human nature; a view that culture is a learning process biased by innate properties of the sensory system and the brain. A view of genes and culture that have coevolved. The new confidence in the unity of knowledge, Wilson believes, has been supported by four developments. The first is cognitive neuroscience as the physical mapping of mental process. A second is human behavior genetics. A third is evolutionary biology, including sociobiology which attempts to reconstruct the evolution of brain and mind. The last is environmental science as a consideration of the physical environment to which humanity is genetically and culturally adapted.[14]

In considering these contributions, Wilson places key emphasis upon neuroscience. What this contributes, he believes, is the understanding of emotions that operate through physiological processes. An example of this process is the human smile, which appears in infants, that invokes affection from adults, and reinforces bonding between a parent and child. Phobias are also powerful enough to engage the autonomic nervous system in panic, nausea, etc. Polar vision can be tracked all the way from genes to neurons. Incest avoidance also springs from hereditary epigenetic rules. The Darwinian advantage of incest avoidance, Wilson contends, is overwhelming. For the mortality rate among children born of incest is about twice that of outbred children, and among those who survive, genetic defects such as dwarfish, hearing deformities, and severe mental retardation are ten times more common.[15]

Wilson recognizes that the concept of "consilience" he is defending may be confused as being reductionist and therefore not suitable to explaining the complexities of human life. But Wilson contends that reductionism is nonetheless "the driving wedge of the natural sciences. A reductionist analysis is one that proceeds from more complex and specific phenomena to more underlying phenomena that are less complex and specific." Biochemistry and molecular biology, for example, have given us a clear view of living cells, and they are now yielding other knowledge in regard to mental processes. Both are relevant to human behavior, and there is no reason why all the social sciences and the humanities must be resistant to this apaproach.[16]

In Wilson's view "epigenetic rules" shape the development of the mind and shape interaction through all categories of behavior. "Culture is created by the communal mind, this view holds, and each mind is in turn the product of the genetically structured human nervous system and brain. Genes and culture are therefore inseparably linked."[17] But Wilson emphasizes this does not mean particular forms of culture are genetically determined. Certain cultural norms, he acknowledges, can survive and reproduce better than others; even if guided by the same epigenetic rules of competing cultures. But the connection between genes and culture is never broken and "culture can never have a life entirely on its own."[18]

III

A central component of Wilson's sociobiology is that the concept of a "fitness of human nature" can be formulated as the product of gene-culture co-evolution. Genes in interaction with the physical environment bias the evolution of culture. But this is only "half the circle." The other half is what culture does to the gene.[19] What is thus unique about human evolution, as opposed to say the chimpanzee, then, is that a large part of the environment shaping it has been cultural. Wilson, thus, emphasizes that this contention should not then be confused with genetic determinism. Genes do not specify conventions such as totemism, elder councils, or religious ceremonies. Epigenetic rules only predispose people to develop such conventions. Wilson believes this includes several categories of human evolution. One is *kin selection* which is especially important in the origins of altruistic behavior, illustrated, for example, by two sisters: one who sacrifices her life or remains childless in order to help her sister. As a result, the sister raises more than twice as many children as she would have otherwise. A second illustration is *parental investment*: the behavior toward offspring that increases their fitness, fitness of the latter at the cost of the parents' ability to invest in other offspring. *Mating strategy* involves emphasis upon the greater state of women in sexual activity due to time for reproduction and heavy investment required for child rearing, child care, tendency of males to be more sexually aggressive, while women are more coy and selective. *Status* is a further feature of human fitness pertaining to rank, class, or wealth. *Territorial expansion* and *defense* have been cultural universals in traditional tribal groups and the modern state. Territoriality arises during evolution where there is a shortage of food, water, nest sites, and territory available for these resources. *Contractual agreement* pervades social behavior. All mammals, including humans, form societies based on a conjunction of selfish interests, devoting their energies to their own welfare and that of close kin. Wilson believes that *incest* provides the fullest test of a genetic fitness hypothesis. Sexual activity in all society is relatively uncommon between parent and offspring, and long term unions made with the consensual purpose of having such children are almost nonexistent.[20]

It is not difficult to understand how Wilson's view of behavior patterns associated with natural fitness could be subject to the charge of a conservative ideological bias in the legacy of social Darwinism, a critique that was particularly significant in the 1970s coming from a radical left group, *The Science for the People Committee*.[21] But there are two objections to their charge. One is the point made by Wilson and Lumsden in their book *Promethean Fire*. What is well meaning in such a critique, they acknowledge, is the belief that any type of genetic determinism offers support for racism, sexism, and the status quo. But the flaw in this argument, they contend, is the assumption that scientific discovery should be judged on its possible political consequences rather than what is true or false. What this charge misses is that if society members "regard racism or any other kind of injustice as undesirable, whatever the cause, its ravages can be mitigated. Acknowledgment of the hereditary tendency and biological mechanism, thus, becomes a valuable part of political reform rather than an obstruction to it in the same way that information about the sickle-cell anemia and hemophilia is essential to the diagnosis and treatment of these disorders."[22]

But the more important point to be emphasized is simply that there is nothing in Wilson's view of characteristics of natural fitness, as outlined above, that warrants the charge he is seeking to provide a defense for a conservative political ideology. If it is possible for Wilson to defend a genetic basis for gender difference in that men are more physically aggressive or "venturesome" than women, it would suggest that the universal existence of sexual division of labor is not entirely an accident of cultural evolution. But this contention is not incompatible with the full recognition of the struggle for women's rights that is now spreading throughout the world. For it only indicates the need to consider varying possibilities. We can seek to exaggerate such difference, as well as to eliminate them; to simply emphasize equality of opportunity and take no further action. "The evidence of biological constraints," Wilson believes, "cannot alone prescribe an ideal course of action, but it can help us define the options and assess the price of each."[23]

If aggression and territoriality may have a genetic basis, Wilson contends, the learning of rules of violent aggression are largely obsolete. "We are no longer hunter gatherers who settle disputes with spears, arrows, and stone axes." This does not mean rules of aggression can be

banished. We can only work around them, work the possibility of finding a more durable foundation for peace in which these objects of human
nature can be "gently hobbled into interests of human welfare."[24] There
is thus nothing in Wilson's view of a Darwinian natural fitness that
warrants the charge that it provides a conservative defense of existing
structures of power and domination, but that it is rather an endorsement
of a pragmatic orientation to public policies directive to the reconciling
of conflicting claims and interests.

IV

From what is indicated above, in Wilson's view of characteristics of
natural fitness, if they have a genetic basis, such characteristics do not
dictate policy choices but only what needs to be taken into account in
regard to variable possibilities and options. It is here that the problem of
ethical evaluation emerges.

Wilson's view is that human altruism has a genetic basis. But it
should be emphasized that the form and intensity of altruism are to a
large extent culturally determined. There is an underlying emotion
manifested in virtually all human societies that has evolved through
genes. This entails two different forms. One is a "hard core" altruism
where there is no expectation of reward and restricted to closest rela-
tives. A second is a "soft core" altruism which is ultimately "selfish,"
and where there is an expectation of reciprocity.[25] Wilson is fully aware
that his contention is subject to the charge that it is an example of the
"naturalistic fallacy" that one cannot derive a moral assertion from
empirical observation. But, in Wilson's view, the origin of ethical reasoning can be objectively based. For the individual is predisposed to
make certain choices. "Strong inner feeling and historical experience
cause concern to be preferred; we have experienced them, and weighed
their consequences, and agreed to conform with codes that express them.
Let *us take the* oath upon the codes, *invest out personal honor* in them,
and suffer punishment for their violation."[26] An empirical view, Wilson
contends, concedes that moral codes are devised to conform to some
drives of human nature and to suppress others. "*Ought* is not the translation of human nature but of the public will, which can be made increas-
ingly wise and stable through the understanding of the needs and pitfalls
of human nature."[27] Wilson believes that the view of ethical codes
arising from the interplay of biology and culture is a renewal of a tradition of moral sentiment developed in the eighteenth century by British
empiricists such as Francis Hutchison, David Hume, and Adam Smith.[28]

The best way of envisioning the meaning and the concept of reciprocal altruism, he contends, is provided by game theory, particularly in the solution postulated by the "prisoners' dilemmas." This is where two gang members have been arrested for murder and are being questioned separately, but where the evidence against them is not decisive. The first gang member believes that if he turns state's evidence, he will be granted immunity, and his partner will be sentenced to life in prison. But he is aware his partner has the same option. Since this is the dilemma, they agree in advance to remain silent with the hope that by doing so, they will be convicted on a lesser charge, or escape punishment altogether. What this entails, then, is that "honesty does exist even among thieves." Such a code would be the same as that of a captive soldier obliged to give only name, rank, and serial number. Such dilemmas, Wilson believes, are true of everyday situations "whether the payoff is money, status, access, comfort, and health—most of these proximate rewards are converted into the universal bottom line of Darwin genetic fitness: greater longevity, and a secure growing family."[29]

It would be important to emphasize that if Wilson perceives a genetic altruism as a concern for an individual cost-benefit calculation, he believes it can be congruent with the needs and interests of humanity as a whole, and the requirements of an environmental ethic. Wilson recognizes what has been an "exceptionalist" view in which the species exists apart from the world and holds dominion over it, exempt from the laws of ecology. But a competing image is a "naturalist self image," recognizing how the human body and mind are adapted to this world; a "habitat selection" prescribed by our genes that is the course of the survival of humanity.[30] Such a contention is thus congruent with an environmental ethic concerned with the basis of human overpopulation, and the destruction of the planet's resources. Such an environmental ethic thus stands against the market-based global economy where the main players are the military, the most powerful and "in spite of a great deal of rhetoric largely indifferent to the suffering of others."[31] For what we confront is a poverty few people are aware of, in which little more than one-fifth of the world population have incomes under one U.S. dollar a day, and where each year between 13 to 18 million, mostly children, die of starvation.[32]

It is for the same reason that Wilson believes that an environmental realism is incompatible with the "myopia" of the classical economist, the "insular nature" of neoclassical economic theory, with its "elegant coherent specimens of applied mathematics," that generally ignores the human behavior understood by contemporary psychology and biology.[33]

Wilson thus believes that the evidence of evolutionary biology is a
defense of a conservation ethics. "The legacy of the Enlightenment is the
belief that entirely on our own, we can know, and in knowing, understand, and in understanding, choose wisely. . . . Human autonomy having
thus been recognized, we should now feel disposed to reflect on where
we wish to go."[34]

V

It was the intent above to show that there is nothing in Wilson's sociobiology that warrants critical objections that he provides support for
current forms of conservative ideology, and certainly nothing that can be
fairly accused of a continuity with social Darwinism. Yet this is not to
deny several serious difficulties in his version of a Darwinian renewal.
One is an obvious ambiguity in Wilson's view of the need for a consilience of Enlightenment science with social sciences and humanities and
how such consilience can be an essential approach to clarification of a
framework for Darwinian renewal. It was seen that Wilson celebrates the
Enlightenment scientific tradition in terms given classical expression in
the materialism of Bacon, Newton, and Galileo. But it should be noted
that Wilson is cognizant of the inadequacy of a radical reductionism. He
doubts that higher levels of organic development can simply be reducible
to lower levels, and he emphasizes upon the necessity of *affirming* a
concept of free will of human intentionality in ethical evaluation. The
body and brain, he concedes, involves "discordant patterns the unaided
conscience cannot even imagine." Thus there cannot be a determinism of
human thought, at least not in obedience to the causation "in the way
physical forces describe the notion of bodies and the atomic assembly of
molecules. Because the individual mind cannot be fully known and
predicted, the self can go on passionately believing in our own free
will."[35]

But Wilson does not clarify this contention as a basis for showing
how the contribution of neuroscience, that he believes is the key to
understanding human evolution, can avoid the inadequacy of a reductive
materialism. What is only suggestive in Wilson's contention requires a
clarification effectively provided by John Searle. Searle is fully cognizant that conscious states are caused by brain processes. But conscience
consists of inner subjective states or "a first person ontology" that cannot be reduced to a "third person" phenomena such as heat, liquidity, or
solidarity. Searle emphasizes the centrality of "speech acts," with an
"illocutionary" component having to do with what is *assertive* in regard

to a state of affairs in the world; a *commissive* component in regard to undertaking a course of action; an *expressive* component as apology, welcome, condolence; a *declaration* to bringing about a change in the world. Central to the illocutionary component of human intentionality is the role of linguistic expression. The role of language is the basis for differentiation of human from animal life. Animals have a primitive intentionality, belief, perception, desires. But once the child acquires language, the capacities for intentionality increase.[36]

From the above comments Wilson makes in regard to the importance of sustaining free will and intentionality, there is no reason why he could not readily endorse the view of Searle. But if so, a difficulty must be confronted. This entails what seems incoherent in Wilson's contention that higher levels of organic life cannot simply be reduced to lower levels, as well as his cognizance of the necessity of defending human free will and intentionality, while yet celebrating a version of the Enlightenment inspired by Newton and Galileo that entails a reductive materialism. Overcoming this contradiction does not require a repudiation of what Wilson rightly believes to be the authentic achievement of the Enlightenment heritage in regard to developments in physics, chemistry, and biology, and his view of the need for consilience of that achievement with the social sciences and humanities. But it does require the effort to show how the Enlightenment achievement can be salvaged from the distortion of a reductive materialism in terms that could sustain a constructive continuity with a more Aristotelian naturalism that can avoid both the inadequacies of a mind-body dualism as well as a reductive materialism. This possibility will be given further elaboration in the concluding chapter of this study.

A second serious objection to Wilson's sociobiology is simply the inadequacy of a "cost-benefit calculus" as an approach to moral evaluation. Here Wilson has been subject to a powerful critique by Mary Midgley. Midgley is fully affirmative of the contemporary trend toward genetic-culture interaction, the distinguishing marks of man (speech, rationality, culture) are not opposed to nature—"but continuous with and growing out of it." But Midgley strongly objects to Wilson's view that moral evaluation can be reducible to a genetic altruism. This is not to deny that we do choose, in some senses, what will well please or satisfy us. But this is as true of suicide, alcoholism, or "unprofitable vengeance" as it is of eating and drinking. Thus what we choose may often *attract* us but without *paying* us in any normal sense. A basic distortion in Wilson's view of a "reciprocal altruism," she contends, is in the example he gives of a Darwinian calculus in the rescuing of a drowning man. If such episodes are true, there would be little gain. But if the drowning man reciprocates in the future, both individuals will have benefited. The fallacy of this contention, Midgley believes, is the fact that people do *in fact* quite often rescue strangers *without* checking that these strangers are "husky, loyal, useful allies who will stay around until a counter rescue becomes necessary."[37]

If intelligence had really been the only impelling force, most of the commitments would never have been found necessary. Why affection? Why time-consuming greeting procedure, mutual grooming, dominance and submission displays, territorial boasting, and ritual conflict? Why play? Why kind-idle chatter, love making, sport, laughter, song, dance, story-telling, quarrels, ceremonial mourning, and weeping? Intelligence alone would not generate these ends. It would just calculate means. But these things are done for their own sake; they are part of the activity that goes to make up the life proper to each species. Insofar as there is one impelling force it is sociability. From that comes increasing power of communication which provides the matrix of intelligence.[38]

A further difficulty in Wilson's genetic altruism, from the standpoint of Midgley's analysis, is the one-dimensional qualitative approach. For a conflict of goods is at the heart of facts about human wants and needs, and thus the problematic features of what nature demands. For human needs are multiple and frequently in conflict. "Love clashes with honor, order with freedom, art with friendship, justice with prudence." What this requires, then, is critical balancing of conflicting claims in order to attain full growth, a balancing congruent with Aristotle's doctrine of "virtue as a mean."[39]

What Midgley's critique of Wilson's cost-benefit calculus in moral evaluation entails at the level of individual action can also be applied to the dimension of political action. This is well articulated by Jon Elster's view that democratic suffrage and the welfare state, for example, cannot be based on a version of society as a "joint stock company with citizens cooperating for mutual advantage" for both the ideals as suffrage and the welfare state embody more compelling conceptions "transcending both instrumental and commutative justice." Such ideals are, indeed, a joint venture, but the bond among the members is not simply one of mutual advantage, but also one of "mutual respect and tolerance." What this entails, then, is that reform movements directive to extension of the suffrage and overcoming inequalities are expressive of people's willing-ness and motivation for "putting up with the costs of transition and of experimenting with differing modes of implementing it." Elster thus invokes John Rawls's concept of a distributive justice as a choice of citizens under a "veil of ignorance" in regard to arbitrary features due to inborn disabilities, handicaps, or personal qualities of endowment[40]

A final objection to Wilson's sociobiology is simply that it is not convincing that human altruism as well as traits of natural fitness such as gender difference, territorial defense, social status, etc., have any definite genetic basis. Wilson, of course, is careful to emphasize that what he is contending is a "co-genetic-cultural *interaction*," but he nonetheless insists upon a concept of "epigenetic rules" and that genes "hold culture on a leash."[41] It is this feature of Wilson's sociobiology that is thus subject to the objection that he is *still* subscribing to a form of biological reductionism. It is this objection that has been given influential articulation by Steven Gould and R. C. Lewontin, who are leading exponents of an alternative to Wilson's sociobiology as a Darwinian pluralism that will be considered in the following chapter.

Notes

1. Edward O. Wilson, "Human Decency As Animal," *New York Times Magazine*, October 12, 1975, 39.

2. Edward O. Wilson, *Consilience: The Unity of Knowledge* (New York: Alfred Knopf, 1998), 21.

3. Wilson, *Consilience*, 27.

4. Wilson, *Consilience*, 31.

5. Wilson, *Consilience*, 32.

6. Wilson, *Consilience*, 33.

7. Wilson, *Consilience*, 39.

8. Wilson, *Consilience*, 39.

9. Wilson, *Consilience*, 181.

10. Wilson, *Consilience*, 185.

11. Wilson, *Consilience*, 41.

12. Wilson, *Consilience*, 190.

13. Edward O. Wilson, "Resuming the Enlightenment Quest," *Wilson Quarterly* (winter 1998), 16.

14. Wilson, "Resuming the Enlightenment Quest," 18.

15. Wilson, "Resuming the Enlightenment Quest," 21-23.

16. Wilson, "Resuming the Enlightenment Quest," 25.

17. Wilson, "Resuming the Enlightenment Quest," 23.

18. Wilson, "Resuming the Enlightenment Quest," 24.

19. Wilson, *Consilience*, 165.

20. Wilson, *Consilience*, 180.

21. See Charles Lumsden and Edward O. Wilson, *Promethean Fire: Reflections on the Origin of Mind* (Cambridge, Mass.: Harvard University Press, 1983).

22. Lumsden and Wilson, *Promethean Fire*, 41.

23. Edward O. Wilson, *Human Nature* (Cambridge, Mass.: Harvard University Press, 1978), 134.

24. Wilson, *Human Nature*, 120.

25. Wilson, *Human Nature*, 156.

26. Wilson, *Consilience*, 251.

27. Wilson, *Consilience*, 251.

28. Wilson, *Consilience*, 252.

29. Wilson, *Consilience*, 278.

30. Wilson, *Consilience*, 282.

31. Wilson, *Consilience*, 282.

32. Wilson, *Consilience*, 290.

33. Wilson, *Consilience*, 297.

34. Wilson, *Consilience*, 220.

35. John Searle, *Mind, Language and Society* (New York: Basic Books, 1998), 146-52.

36. Mary Midgley, *Beast of Man: The Roots of Human Nature* (New York: Routledge, 1979), 321.

37. Midgley *Beast of Man*, 127.

38. Midgley *Beast of Man*, 130.

39. Midgley *Beast of Man*, 192.

40. John Elster, "The Possibility of Rational Politics," in *Political Theory Today*, edited by David Held (Stanford, Calif.: Stanford University Press, 1991), 139-40.

41. Lumsden and Wilson, *Promethean Fire*, 60.

Chapter 3

Gould, Lewontin, Rose, and Kamin: Darwinian Pluralism

From what has been indicated in the previous chapters, Darwinian theory of human evolution has given rise to conflicting ethical-political implications. It was the intent of the introduction to show that Darwin's *Descent of Man* is indicative of a continuity with Aristotelian implications: a view of human ethical capacity directive to concern for a common good beyond individual egoism and self-interest. The association of Darwin's theory of evolution with the conservative ideology of nineteenth-century competitive individualism, it was contended, was derivative from Darwin's *Origin of Species* that was an interpretation of animal rather than human evolution. But it was also noted that features of Darwin's *Descent of Man* having racist, class implications could be a source of support for social Darwinist ideology. While developments in social sciences in the early part of the twentieth century, it was noted, led to the repudiation of social Darwinism and the ascendency of culturalism, more contemporary developments, it was seen, have given rise to a striking renewal of Darwinian theory of evolution, but with continuing controversy and disagreement on how it can be most effectively clarified. It will be the intent of this chapter to provide a balanced view of what is constructive as well as unsatisfactory in the concept of a Darwinian pluralism giving leading articulation by Stephan Gould, R. C. Lewontin, Steven Rose, and Leon Kamin. It will be shown that their critiques of the misuse of scientific evidence on behalf of conservative political ideologies provides a powerful reinforcement for the critical reaction to social Darwinism that was noted in chapter 1. They also provide a constructive emphasis upon a "Darwinian pluralism" as an emphasis upon "local adaptation," or what has also been characterized as "punctuated equilibrium." But it will be argued this contention can be appropriated within the more orthodox view of Darwin's natural selection. It will then be argued that while these writers are persuasive that a concept of Darwinian evolution must be seen as simply *interactionism* without being able to specify what is specifically genetic versus what is specifically cultural, they unfairly place Wilson's sociobiology in the legacy of social Darwinism. For this does not take account of how Wilson is able to defend a limited form of genetic causation that can be consistent with ideals of political liberalism. The more serious objection to Gould-Lewontin-Rose-Kamin's concept of a Darwinian pluralism is simply that if they provide a convincing revelation of how forms of biological reductionism have served a conservative political ideology, there is an unresolved confusion and incoherency in what they believe to be ethical/political implications of a Darwinian pluralism. On the one hand their defense of a Darwinian pluralism is simply "open-ended stories" in which "anything is possible." Yet, on the other hand, their powerful critique of all forms of biological reductionism indicates an implicit commitment to a left-wing ideology of equality and social justice, but without any clarification that this is what they perceive to be the implications of a Darwinian pluralism.

I

In his book *The Mismeasure of Man*, Steven Gould declares his intention to demonstrate both the scientific weaknesses and the political context of determinist arguments. But Gould emphasizes that: "I do not intend to contrast evil determinists who stray from the path of scientific objectivity with enlightened anti-determinists who approach data with an open mind and therefore see truth. Rather I criticize the myth that science itself is an objective enterprise, done properly only when scientists shuck the constraints of their culture, and view the world as it really is." Gould contends that he is not, then, saying that biological determinists are "bad scientists" or "always wrong." He is saying, rather, that science must be understood as a social phenomenon, and not the "world of robots programmed to collect information."[1] Science, he believes, is a "socially embedded action" for culture influences what we see and how we see it. This is not, however, to embrace a relativist view that science simply reflects a given social context. There is a factual reality which scientists can learn about: how Galileo, for example, threatened the church's conventional view of society as a "doctrinal stability." But the church, finally, had to come to terms with Galileo's cosmology: "the earth really does revolve about the sun." But the history of science is no less affected by political constraints for two main reasons. One is where the relation of data to social implication is low. Science can thus become a powerful agency for social change. This is well illustrated in the scientific view of race such as Hitler's use of racist arguments for sterilization and racial purification. Second, scientific questions are formulated in such a way that often validate preferences such as data on racial differences in mental capacities that are seen as a "thing in the head." Until this notion was swept aside, "no amount of data could dislodge a strong Western tradition for ordering related items into a progressive chain of being."[2]

Biological determinism, Gould believes, is based on two deep fallacies. One is that because of our recognition of the importance of mentality in our lives, we wish to make distinctions among people in our culture and politics, and we *"reify* intelligence as a unitary thing." A second fallacy is the ranking of the metaphors of "progress" and "gradualism" that leads to a criteria of assigning individuals to their proper status in a single series, and to invariably finding that oppressed and disadvantaged groups (races, classes, or sexes) are thus innately inferior and deserve their status. What this entails, then, is the "Mismeasure of Man."[3] Gould surveys several examples of this development that can be seen as a reinforcement of the survey of social Darwinism considered in chapter 1 that are indicative of how biological determinism has been subservient to conservative political ideologies of time and place. One was the racist ideology fostered by the scientific theory of polygeny and cranometry. All leading scientists of this era, he notes, followed social convention. Louis Agassiz, for example, was an American theorist of polygeny in the nineteenth century who saw a definitive biological criteria for separate races with blacks at the bottom, lest the white race be compromised and diluted; the dangers of mixed and enfeebled people, the necessity for rigid separation among races.[4] Samuel Norton was an exponent of "polygeny" that he believed demonstrated the character deficiencies of racial groups such as the Greenland Esquimaux, Indians, and blacks. Gould believes these scientists are classical examples of the abuse of scientific investigation in the service of cultural prejudices, inconsistency, and shifting criteria, subjectively directed toward a prior prejudice; procedural omissions that seem obvious to us; miscalculation, and convenient omissions.[5] Paul Broca, Gould points out, was one of the nineteenth-century masters of craniology, and an exponent of how scientific analysis was an instrument of racism and sexual prejudices in which conclusions come first that were prejudices of white males of his time, where facts were gathered selectively and manipulated in service of prior conclusions.[6]

A further illustration of the distortions of scientific theory, Gould believes, was the hypothesis about the biological nature of criminal behavior given leading expression by Cesar Lombroso. This entailed the view that criminals have "smaller brains, peculiar speech, use of tatooing, atavistic love of adornment, a tendency to epilepsy." Criminal anthropology, Gould comments, can't be beat as a conservative political argument: "Evil or stupid, or poor, or disenfranchised, or degenerate, people are what they are as a result of their birth. Social institutions reflect nature. Blame (and study) the victim not his environment."[7] But Gould also notes that criminal anthropologists were not necessarily "proto-Fascist," or even always conservative ideologists. Some even turned toward liberalism or even socialist politics, seeing themselves as "scientifically enlightened modernists." They hoped to use modern science as a "cleaning broom to sweep away from jurisprudence the outdated philosophical baggage of free will and unmitigated moral responsibility. They called themselves the 'positive school of criminology,' not because they were so certain, but in reference to the philosophical meaning of empirical and objective rather than speculative."[8]

In the last part of the nineteenth century and early part of the twentieth century, Gould contends, the hereditary theory of IQ became the leading basis for racist prejudice. Alfred Benet was a leading exponent of this development. His theory, Gould contends, was a practical device in identifying mildly retarded and learning disabled children who needed special help and that low scores should not be used to identify children as innately incapable. But American psychologists perverted his intentions as the basis for a hereditary theory of IQ, assuming intelligence was inherited despite variations in the quality of life.[9] It was H. H. Goddard, Gould notes, who popularized Benet's scale in America, regarding the scores as a basis for segregating and curtailing breeding and the further "deterioration of an endangered American stock."[10] Lewis Terman was a leading figure in establishing the hereditary basis of intelligence associated with race and class.[11] R. M. Yerkes promoted the use of mental tests during World War I. Such tests also became the basis for his grading European immigrants in terms of their origins. "The darker people of Southern Europe and the Slavs of Eastern Europe are less intelligent than the fair peoples of Western and Northern Europe. Nordic supremacy is not a jingoistic prejudice. The average Russian has a mental age of 11.34; the Italians 11.01; the Poles 10.74. The Polish joke attained the same legitimacy as the moron joke—indeed they describe the same animal. . . . They lie at the bottom of the scale with an average mental age of 10.41."[12] Such studies, Gould points out, were the basis for quotas that it is estimated barred six million Southern, Central, and Eastern Europeans from the United States between 1924 and the outbreak of World War II.[13]

Gould believes that the most recent manifestation of scientific biological determinism is Richard Hernstein and Charles Murray's *The Bell Curve*. What this entails, he believes, is an "anachronistic social Darwinism," and that its success in "winning such attention must reflect the depressing temper of our time when a mood of slashing social programs can be abetted by an argument that limits as low IQ scores."[14] Gould contends that the central distortion in *The Bell Curve* is the concept of "multiple regression" in which social behavior such as crime and unemployment are weighed against IQ and parental socioeconomic status. By holding socioeconomic status constant, and considering the relationship of the same social behavior to IQ, they find more correlation with IQ than with socioeconomic status. The fundamental fallacy in this correlation, Gould contends, is that their numerous graphs show only the *form* of this relationship; that is, they show the regression curve of their variables against IQ and parental socioeconomic status. "But in violation of all statistical norms that I've ever learned they plot only the regressive curve and do not show what the scatter is of variation around the curve." Thus their graphs show nothing about the strength of the relationship— that is, the amount of variation in social fact are explained by IQ and socioeconomic status.[15] It is Gould's conclusion that *The Bell Curve* is not an academic treatise on social theory and population genetics, but a "manifesto of a conservative ideology . . . the text evokes the dreary and scary drumbeat of claims associated with conservative think tanks— reduction and elimination of welfare, ending of affirmative action in schools and workplaces, cessation of Head Start and other forms of preschool education programs."[16]

II

Gould's view of how forms of biological reduction have served the interests of a conservative political ideology has been effectively reinforced by R. C. Lewontin, Steven Rose, and Leon Kamin in their book *Not in Our Genes*.[17] What this entails, they believe, requires a clarification of the relationship of science to political philosophy. They emphasize that they are not arguing that political philosophy or the social position of the scientist invalidates their claims. What they argue, however, is two distinct claims that must be differentiated in decisions of the world around one. One is the claim influenced by Thomas Kuhn that criteria of normal science can be modified by periods of revolutionary science where existing paradigms are shaken.[18] A second claim of equal importance is the social matrix in which science is embedded; a recognition of the questions asked by the scientist; what explanations are appropriate to the paradigm framed; and criteria of weighing evidence are historically relative. It is from this perspective, they believe, that one can contend that the positivist tradition of scientific knowledge was part of the "objectification of social relationship" in the transition from feudalism to modern capitalist society. What this involved was the determination of a person's status and role in society by his or her relation to objects; and where scientists saw individuals confronting an objective and external nature, rather than their relation to one another, to the state, patrons, and owners of wealth and production. Scientists had done more than simply participate in the objectification of the society, for they had raised the concept of objectification to an absolute good called "scientific objectivity." But the emphasis upon objectivity masked the relation of scientists to each other and with the rest of society. "By denying these relations, scientists make themselves vulnerable to a loss of credibility and legitimacy when the mask slips and the social reality is revealed."[19]

Lewontin, Rose, and Kamin believe that contemporary trends toward biological determinism are the manifestation of a view that individuals are "ontological prior to society," and that the characteristics of individuals are a consequence of their biology. They are convinced that the origin of determinism came with the break from feudal society that centered upon status relationship, the ideology of grace and divine right;, and humanity's relation to nature conceived as one of coexistence rather than domination. A rise of modern bourgeois ideology was marked by the ascendancy of monetary rather than social relationships, the development of large-scale industrial production carried on by wage workers selling their labor power to owners of capital. Workers also became predominately male so that the new order became not merely capitalist, but patriarchal. A further feature of the new economic reality was a "presumptive equality." For the growing bourgeois class, and the rise of the entrepreneur was needed to acquire and dispose of land and property, along with a legal system to redress grievances against nobles, access to political power, and the "supremacy of a parliament of commoners." [20]

Accompanying this development was a new type of science influenced by Bacon: the acquisition of facts about the world and experimental manipulation of those facts that were integral to new theories. Thus, "the new science, like the new capitalism, was a part of the liberation of humanity from the shackles of feudalism and of human ignorance . . . even the most abstract pronouncements of physics such as Newton's law of motion could be seen as arising out of the social needs of an emergent class. Science was thus an integral part of the new dynamic of capital, even though the fuller articulation of the links between them would take another two centuries to develop."[21]

In the view of Lewontin, Rose, and Kamin, it is not surprising that the philosophical principles of the Enlightenment corresponded to the demands of the bourgeois relations in which the idea of freedom and equality provided the revolutionary rhetoric for throwing off the dominance of the church and the aristocracy. "The machine model of the universe gained intellectual hegemony, ceasing to be regarded as merely a metaphor and becoming, instead, the self evident truth about how to look at the world."[22]

What Lewontin, Rose, and Kamin also see to be of central significance in the new bourgeois view of nature was a reductionist principle indebted to Galileo and especially Newton: an atomization of the natural world that had its parallel in the commodity exchange of capitalism. Later developments influenced by Einstein's demonstration of the equivalence of matter and energy "corresponded to an economic reductionism whereby all human activities could be assessed in terms of their equivalence of pounds, shillings, and pence." Thus for bourgeois society "nature and humanity itself has become a source of raw materials to be extracted, an alien force to be controlled, tamed, and exploited in the interest of the newly dominant class. The transition from the precapitalist world of nature could not be more complete."[23]

Lewontin, Rose, and Kamin believe that the beginning of modern biology was provided by Descartes, a "Cartesian machine image" that has come to dominate science—acting as a fundamental metaphor that legitimized the bourgeois worldview. By the nineteenth century, a radically reductionist program became prominent in the views of leading physiologists and biological chemists.[24] Coming out of this was a further step toward the nature and origin of life itself. It was Darwin who formulated the mechanism for evolution, changing the terms of natural selection. The concept of natural selection became the basis for a neuropsychology as a view of the brain in controlling behavior, and a basis for theorizing, for example, that criminals could be identified as having certain physiological features.[25]

Lewontin, Rose, and Kamin place strong emphasis upon how the development of a biological determinism became the support for the emerging ideology of bourgeois revolution in which what was initially an idea of freedom and equality that served as the basis for revolutionary class struggle become the legitimation of the ideology of the class in power. The rise of biological determinism thus became a defense of inequality seen as the result of difference in intrinsic merits of individuals. Lewontin, Rose, and Kamin believe, like Gould, that IQ tests have been the basis for legitimizing inequalities, used in the United States and England as a basis for shunting vast numbers of working-class and minority children into inferior and dead-end educational tracks. In their view, differences in IQ are not due to genetic factors, but rather to family background and socioeconomic status.[26] The same is true, they believe, in regard to scientific claims to gender differences. Exponents of gender difference seek to establish that differences in power between men and women are due to hormones, making males more aggressive, females less aggressive. What this entails is tracing complex social interest to biological causes that ignore the wide range of economic and cultural determinants in explaining individual success.[27] Claims for the evolution of patriarchal theory have been derived from historical evidence of food gathering in early human society that gave dominance to males and a division of labor. But it is their contention that the real division of labor in much of recorded history does not require biological explanation. "Nothing is added to our understanding of the phenomena, or of its persistence, by postulating genes for this or that aspect of social behavior. If patriarchy can take . . . any form from baby kissing to crusading, the leash on which genes hold culture . . . must be so long, so capable of being twisted and turned in any direction, that to speculate upon the genetic limits to the possible forms of relations between men and women become scientifically and predictably useless; it can serve only an ideological interest."[28]

The reductionist argument in regard to male dominance, they believe, involves several mistakes. One is the insupportable labeling of behavior as being natural where it is, in fact, simply power relations of male and female emerging at particular times in history. A second distinction is due to the limited nature of the observers accounting for what is happening in social interaction. This involves, for example, the distortion in what ethologists simply take for granted that the male is the main actor; the heterosexual procreative is the only form to be considered; and that the task of the female is only to be receptive. Third, generalizations about patterns of behavior are derived from only a small number of observations that do not take into account widely varying habits and diverse cultural and social phenomena.[29] This distortion is accompanied by the powerful Freudian tradition in psychoanalysis in which differences in the behavior between the sexes is seen as lying (if not in the brain), in the "ineluctable biology of the genitalia." This trend also involves the fallacy of reducing social phenomena to biological determinants that neglect explanation of diverse cultural and social phenomena.[30]

Lewontin, Rose, and Kamin also show how a reductionist biology is apparent in the politicization of psychiatry. This was exemplified in the Soviet Union in the diagnosis of political protest as the expression of mental disturbances.[31] But this type of biological reductionism was also manifested in what was seen to be the role of brain disease in riots and urban violence in the United States in the 1960s.[32] They also point to efforts to establish a genetic basis for "hyperactive child syndrome," which they cite as a further distortion in reductionism in the assumption of a discord caused by a single malfunctioning biochemical substance or genre.

Cures for widespread social distress and individual existential despair in advanced industrial, patriarchal, capitalist, or so-called socialist societies cannot be found merely by manipulating the biology of the individual member of that society. Yet the nature of the society in which we live profoundly affects our biology as well as our behavior. In a healthier and more socially just society, even though pain, illness, and death will never be eliminated, our own individual biologies will nonetheless be different and healthier.[33]

Lewontin, Rose, and Kamin also mount a strong critique on the "politicization of psychiatry" in which biological determinist arguments, such as brain dysfunction, are employed in the explanation for social ills such as poverty, unemployment, violence on the streets, etc. The diagnosis and treatment of schizophrenia, they contend, is a "paradigm of a determinist mode of thinking," where the cause is attributed to a particular molecule of gene. In their analysis of this development, Lewontin, Rose, and Kamin seek to expose the theoretical and empirical distortions in regard to supposedly biological determinants. This is not to deny that there is nothing relevant to be said about the biology of the disorder, or to deny that schizophrenia exists, but rather that this has been made more difficult by "extraordinary latitude available in diagnostic criteria."[34] They are not saying they are endorsing Foucault's view that categories of psychological disorders are simply a historical invention expressing power relations in society and in families. It is rather to argue that a theory of schizophrenia must understand that it is about social and cultural environment that pushes some categories of people toward manifesting schizophrenic symptoms. "What is necessary, then, is not a biological or cultural determinism, but rather a more dialectical understanding of the relationship between the biological and the social."[35]

It is important, finally, to emphasize Lewontin, Rose, and Kamin's view that Wilson's sociobiology is the most recent manifestation of a biological determinism that serves the interest of a conservative political ideology. Such an implication, they believe, is evident in his biological explanation of such cultural manifestations as religion, ethics, tribalisms, warfare, genocide, cooperation, competition, and entrepreneurship. Sociobiology, they contend, is a reductionist, biological determinist explanation of human existence. "If men dominate women it is because they must." If employers exploit wages, it is because evolution has provided genes for entrepreneurial activity. If we kill each other, it is due to genes relating to territoriality, xenophobia, tribalism, and aggression.[36]

Lewontin, Rose, and Kamin believe that sociobiology is part of the legacy of "pop ethology" influenced by Audrey, *The Territorial Imperative*; Lorenz, *On Aggression*; Tiger and Fox, *The Imperial Animal*; Morris, *The Naked Ape*, in whose books one finds a view of humans as being by nature territorial and aggressive, a Hobbesian view of "war of all against all." The influence of Hobbes comes through Darwin and social Darwinism as an emphasis upon the struggle between existence and competitive struggle between organisms.[37] Sociobiology, they contend, treats categories such as aggression, tribalism, and territoriality as natural dispositions without realizing that they are historically and ideologically conditioned constructions.[38] There are several specific errors, they contend, that are made by sociobiologists. One is simply the error of "arbitrary agglomeration." This has to do with how an organism is to be divided into parts in order to understand its evolution. This can be illustrated in regard to the evolution of the hand, which is tied in evolution to other parts of the body, and until these are understood, it is by no means sure that it is the unit of phenotypic description.[39]

The second error is one of "reification," taking concepts that have been created in the understanding of human experience and endowing them with a life of their own. "Just as the Greeks thought that those figments of the imagination, the gods, could reproduce and vanquish each other in battle so sociobiologists think that religion can be inherited and increased in frequency by natural selection in the struggle to exist."[40] A third error is the use of metaphors that are taken for real identity. This is illustrated in the sociobiologist's use of "royalty and slavery in ants from nineteenth-century entomology. Aggression, warfare, etc.

are re-applied to nonhuman animals and human manifestations come to be seen as special, perhaps more developed cases." A fourth error, related to the use of metaphors is the "conflation of different phenomena under the same rubric." This is illustrated in sociobiologists' view that "political aggression is a collective manifestation of individual feelings which fails to see that warfare in state organized societies is a calculated political phenomenon expressive of those in power in society for political and economic gain and little to do with prior individual feelings of aggression."[41]

Lewontin, Rose, and Kamin are convinced that the most fundamental fallacy of sociobiology, finally, is the central assertion that human behavior is coded in the genes. What is unsatisfactory in the model of gene control is that the traits of an organism or its phenotype are not determined by genes in isolation, but are the product of the interaction of genes and environment. Sociobiologists, such as Wilson, try to escape this accusation that they are naive genetic determinists by recognizing the importance of environmental influences, while yet wishing to insist "genes hold culture on a leash." But Lewontin, Rose, and Kamin believe that this qualification simply cannot be framed in the language of genetics in a way that has any technical meaning. In their view, there is simply no evidence of a heritable basis in human traits such as introversion, extroversion, neuroticism, dominance, and schizophrenia. In their view, "no one has been able to relate any aspect of human social behavior to any particular gene or set of genes, and no one has ever suggested an experimental plan for doing so. Thus all statements about the genetic basis of human social traits are necessarily pure speculation, no matter how positive they seem to be."[42]

III

What has been presented above has been an outline of Gould-Lewontin-Rose-Kamin's critique of what they perceive to be distortions in forms of biological reductionism that they believe have been supportive of conservative political ideologies. It is now necessary to outline what they believe can be bases of a "Darwinian pluralism" as a constructive alternative.

It should be noted that Gould and Miles Eldredge have been noted for development of a theory of "punctuated equilibrium" in 1972, a view that species arise in small isolated populations rather than by slow change in large central populations.[43] But it should be emphasized that Gould does not see this to be a rejection of Darwinism, for Darwin's world view contains a central component that has a focus upon the individual as a primary evolutionary agent; his identification of natural selection as the mechanism of adaptation; and his belief in the gradual nature of evolutionary change. But central to Gould's contention is that Darwin is a pluralist. He did give overwhelming importance to natural selection, but he did not reject the influence of other factors.[44] Gould puts into question the claim of sociobiologists that all major patterns of social behavior must be adaptive as the product of natural selection. Gould is skeptical of the view that genes are "discrete and divisable particles, using the traits that they build in organisms as weapons for personal propagation." It is Gould's view that individuals are not "decomposed into independent bits of genetic coding. For the bits have no meaning outside the milieu of their body, and they do not directly code behavior, are not rigidly built by battling genes; they need to be adaptive."[45] Gould does not believe that the determinist views of molecular biology are definitive. We have no final answer, but he predicts the triumph of Darwinian pluralism: "Natural selection will turn out to be far more important than some evolutionist imagines but it will not be omnipotent as some sociobiologists seem to maintain. In fact, I suspect that Darwin's natural selection based on genetic variation has rather little to do with the very behaviors now so ardently cited in its support."[46] Gould hopes that the pluralist spirit of Darwin will thus permeate evolutionary thought. He believes that paleontology has adopted a gradualist orthodoxy at the price of ignoring other fossil records that now vindicate this contention. Natural selection, he believes, can encompass rapid change by specialization in small populations as well as the conventional and immeasurably slow transformation of entire lineages. Gould affirms Aristotle's concept of a "golden mean": a view that "nature is so wondrously complex, and varied, that almost anything possible does happen."[47]

What Gould means by Darwinian pluralism as an alternative to Wilson's sociobiology is well illustrated in his critique of Wilson's "genetic altruism." It is adaptation, he emphasizes, that is the hallmark of Darwinian processes. But this is not to be seen as an argument for genetic control. This can be illustrated, he believes, in social practices of Eskimos where, in case of shortage of goods, aging parents must be left behind to die rather than endanger the survival of the entire family. But such a sacrifice is not evidence of an "altruistic gene," but what is adaptive. "The sacrifice of grandparents is an adaptive but non-genetic cultural trait. Families with no tradition for sacrifice do not survive for many generations. In other families, sacrifice is celebrated in song and story; aged grandparents who stay behind become great heroes of the clan. Children are socialized from their earliest memories to a belief in the glory and honor of such sacrifices.[48] Gould does not doubt that altruism exists in human society, but that there is no evidence it has a genetic basis, and that it can be inculcated equally by learning. But it would be important to emphasize that Gould is not denying the relevance of biology to human nature. The issue is not universal biology versus human uniqueness, but "biological potentiality versus biological determinism."[49]

Gould's emphasis upon a Darwinian pluralism is also central to the contention of Lewontin, Rose, and Kamin in their rejection of the dichotomy between nature versus nurture that has plagued psychology and sociology since the nineteenth century. In their view there is no human behavior built on genes that cannot be modified and shaped by social conditioning. Lewontin, Rose, and Kamin cite Marx's thesis on Fuerbach as a reinforcement of this contention, namely, "the materialist doctrine that men are the product of circumstances and upbringing, and that therefore changed men are products of other circumstances and changed upbringing, forgets that it is men that change circumstances and that the educator himself needs educating."[50] Lewontin, Rose, and Kamin are not denying that social life is related to biology, but that a pluralist response to biological determinism must be interactionism.[51] What is unique in the contribution of Darwin is a method of trial and error, challenges and responses, in which organisms, societies of species "confront problems set for them by an external nature, independent of their own existence, and they respond by trying various solutions until one is found that fits."[52]

The relationship between biology and society, they believe, must be restricted to a concept of a "dialectical" explanation as opposed to reductionism, avoiding any assignment of weights for different partial causes. What is necessary, rather, is the view that "parts and whole codetermine each other.[53] In his book, *Biology and Ideology*, R. C. Lewontin contends that just as there is no organism without an environment, there is no environment without an organism. "They create them. They construct their own environment out of bits and pieces of the physical and biological world, and they do so by their own activities."[54] In viewing the behavior of a bird, for example, we can see that it eats insects part of the year, but switches to nuts when insects are no longer available; flying south in the winter and coming back in the summer; and when it forages for food it tends to stay in the higher branches. "Every word uttered by the ecologist as describing the environment of a bird will be a description of the life activity of the bird. That process of description reflects the fact that the ecologist has learned what the environment of the bird is by watching birds."[55] Lewontin contends that we must thus replace the "adaptivist" view of life with a "constructionist" view. This would emphasize that the environment of the organism is constantly being remade during the life of those living beings. A rational environmental movement, he also contends, must abandon the unfounded commitment to a harmonious and balanced world, and turn to the real question of how people want to live and arrange their lives. We cannot accept a view of sociobiology that human beings have limitations coded in their genes, whether selfish, aggressive, xenophobic, family oriented, etc. "Social organization does not reflect the limitation of individual biological beings, but is their *negations*. This is not to deny that we are natural, material biological objects developing under the influence of the interaction of genes with the external world; the fact, for example, of our size and our having a certain nervous system. But it is our consciousness that creates our environment, its history, and the direction of the future.

Our DNA has a powerful influence on our anatomy and physiologies. In particular, it makes possible the complex brain that characterizes human beings. But having made that brain possible, the genes have made possible human nature, a social nature whose limitation and possible shapes we do not know except insofar as we know what human consciousness has already made possible. In Simone de Beauvoir's clever but deep apothegism, a human being is "l'être dont l'être est de n'être pas," the being whose essence is not having an essence."[56]

Gould-Lewontin, Rose, and Kamin agree, therefore, upon a framework for a Darwinian pluralism beyond reductionism, emphasizing an interaction or interpenetration of genetic and environment, or a more dialectical approach. The political implication of their position becomes apparent in what they see to be Marx's eleventh thesis on Fuerbach: "the philosophers have only interpreted the world in various ways; the point however is to change it."[57]

IV

It is finally necessary to put into balanced perspective both the constructive contribution of Gould-Lewontin, as well as its inadequacies as a framework for Darwinian renewal and its ethical political implications. Gould, Lewontin, Rose, and Kamin clearly provide a powerful critique of the forms of biological reductionism that have been subservient to conservative political ideologies in the traditions associated with nineteenth century social Darwinism noted in chapter 1. But it was noted that Gould does not believe natural selection to be an *exclusive* agent of evolutionary change and that Darwin's view was pluralistic. Gould and Eldredge, it should also be noted, have been leading exponents of what this entails that has been characterized as "punctuated equilibrium" as an emphasis upon random variation. The importance of Gould's contention, then, is what it provides, namely, that it is a constructive critique of biological reductionism without believing this requires a repudiation of Darwinian orthodoxy. As Michael Ruse points out, the concept of punctuated equilibrium does not negate Darwinian orthodoxy. What one will find, rather, is that some of its central insights will be incorporated into existing theory.[58]

But what is unsatisfactory in Gould-Lewontin-Rose-Kamin's formulation of a Darwinian renewal appears at two points. One is simply that it entails an unfair identification of Wilson's sociobiology with past forms of biological reductionism associated with conservative political ideologies and their justification of sexism, racism, and class prejudices. It was seen in the previous chapter that Wilson and Lumsden argue persuasively that some of the critics of the role of biological factors in human behavior are, in effect, calling for repression of scientific research on grounds of how it *might* be used. What this neglects, for example, is a cognizance of the fact that a knowledge of the hereditary basis of sickle cell anemia and hemophilia is essential to diagnosis and treatment of these disorders.[59] What is more to the point is simply that what Wilson perceives to be genetic components in behavior patterns of natural fitness in regard to gender difference, territorial defense, etc. does not lead him to believe this is supportive of a conservative political ideology. In regard to gender difference, for example, he notes that girls seem more disposed to be more social and less physical or adventurous, a greater investment in sexuality. But Wilson is fully cognizant of enormous variations and if there are slight biological differences, such differences are not an obstacle to the contemporary context of concern for women's rights that could be supportive of either the effort to eliminate such differences or simply to emphasize improving equality of opportunity.[60] While Wilson believes a concept of natural fitness is indicative of genetic dispositions toward territorial defense or aggression, he fully acknowledges that we are no "longer hunter-gatherers who settle disputes with stones, axes, and spears." If there is still some tendency for aggression, we are then able to work around this with the full possibility of settling disputes without resorting to violence.[61] The most important point to be emphasized, finally, is that Wilson clearly believes sociobi-ology is consistent with the ethical premises of liberalism and not what Gould-Lewontin, Rose, and Kamin believe to be a conservative ideol-ogy. This is most apparent, it was seen, in his view that contemporary global problems of poverty, overpopulation, and environmental pollution are being exacerbated by capitalist priorities and the myopia of classical economic theory: Wilson's strong endorsement of the necessity of an environmental ethic.[62] This is not to deny what Gould, Lewontin, Rose, and Kamin rightly perceive as an untenable genetic determinism in regard to Wilson's concept of natural fitness and in his concept of a genetic altruism. But here the question is whether *their own* view of a Darwinisn pluralism is a satisfactory alternative to Wilson in regard to ethical-political implications. It would be possible to contend, of course, that a Darwinian pluralism has no ideological implication; for, as noted above, it entails what are simply "adaptive stories" in which "anything is possible," But, from what has been previously indicated, Gould-Lewontin, Rose, and Kamin are not critiquing biological reductionism from a posture of impartial, disinterested neutrality, but from a left-wing political commitment with which they have been commonly associated. How this commitment is related to their concept of Darwinian pluralism is apparent in their affirmation of its congruence with Marx's eleventh thesis on Fuerbach: "The philosophers have interpreted the world in various ways; the point, however, is how to change it." But what is inherent in a Marxist analysis is also an emphasis that such change must be directive to overcoming structures of power and oppression in order to achieve human goals of equality and social justice. One might assume, then, that Gould-Lewontin, Rose, and Kamin do believe that *this is* an ethical-political implication of their celebration of a Darwinian plural-ism, but no where do they clarify this implication.

Notes

1. Stephen Gould, *The Mismeasure of Man* (New York: W. W. Norton, 1981), 53.
2. Gould, *The Mismeasure of Man*, 55.
3. Gould, *The Mismeasure of Man*, 57.
4. Gould, *The Mismeasure of Man*, 81.
5. Gould, *The Mismeasure of Man*, 88.
6. Gould, *The Mismeasure of Man*, 100.
7. Gould, *The Mismeasure of Man*, 117.
8. Gould, *The Mismeasure of Man*, 166.
9. Gould, *The Mismeasure of Man*, 170.

10. Gould, *The Mismeasure of Man*, 187.
11. Gould, *The Mismeasure of Man*, 189.
12. Gould, *The Mismeasure of Man*, 219.
13. Gould, *The Mismeasure of Man*, 227.
14. Gould, *The Mismeasure of Man*, 263.
15. Gould, *The Mismeasure of Man*, 367.
16. Gould, *The Mismeasure of Man*, 324.
17. Gould, *The Mismeasure of Man*, 276.
18. R. C. Lewontin, Steven Rose, and Leon Kamin, *Not in Our Genes: Biology, Ideology and Human Nature* (New York: Pantheon Books, 1984).
19. Lewontin et al., *Not in Our Genes*, 32.
20. Lewontin et al., *Not in Our Genes*, 33.
21. Lewontin et al., *Not in Our Genes*, 41.
22. Lewontin et al., *Not in Our Genes*, 42.
23. Lewontin et al., *Not in Our Genes*, 45.
24. Lewontin et al., *Not in Our Genes*, 45.
25. Lewontin et al., *Not in Our Genes*, 53.
26. Lewontin et al., *Not in Our Genes*, 94.
27. Lewontin et al., *Not in Our Genes*, 155.
28. Lewontin et al., *Not in Our Genes*, 157.
29. Lewontin et al., *Not in Our Genes*, 159.
30. Lewontin et al., *Not in Our Genes*, 161.
31. Lewontin et al., *Not in Our Genes*, 165.
32. Lewontin et al., *Not in Our Genes*, 169.
33. Lewontin et al., *Not in Our Genes*, 195.
34. Lewontin et al., *Not in Our Genes*, 228.
35. Lewontin et al., *Not in Our Genes*, 231.
36. Lewontin et al., *Not in Our Genes*, 237.
37. Lewontin et al., *Not in Our Genes*, 241.
38. Lewontin et al., *Not in Our Genes*, 247.
39. Lewontin et al., *Not in Our Genes*, 247.
40. Lewontin et al., *Not in Our Genes*, 249.
41. Lewontin et al., *Not in Our Genes*, 251.
42. Lewontin et al., *Not in Our Genes*, 251.
43. N. Eldredge and Steven Gould, "Punctuated Equilibrium: An Alternative to Phyletic Gradualism," in *Models in Paleobiology*, edited by J. S. Schopf (San Francisco: Freeman & Cooper, 1972), 82–115.
44. Stephen J. Gould, *Ever Since Darwin: Reflections in Natural History* (New York: W. W. Norton, 1977), 270.
45. Gould, *Ever Since Darwin*, 270.
46. Gould, *Ever Since Darwin*, 270.
47. Gould, *Ever Since Darwin*, 271.
48. Gould, *Ever Since Darwin*, 256.
49. Gould, *Ever Since Darwin*, 252.

50. Lewontin et al., *Not in Our Genes*, 267.

51. Lewontin et al., *Not in Our Genes*, 268.

52. Lewontin et al., *Not in Our Genes*, 271.

53. Lewontin et al., *Not in Our Genes*, 282.

54. R. C. Lewontin, *Biology As Ideology* (New York: Harper, 1992), 109.

55. Lewontin, *Biology As Ideology*, 110.

56. Lewontin, *Biology As Ideology*, 123.

57. Lewontin et al., *Not in Our Genes*, 277; Gould, *Urchin in the Storm: Essays about Books and Ideas* (New York: W. W. Norton, 1997), 154.

58. Michael Ruse, *Darwinism Defended: A Guide to the Evolution of Controversies* (London: Addison-Wesley Publishing, 1982), 225.

59. Charles Lumsden and Edward O. Wilson, *Promethean Fire: Reflections on the Origin of Mind* (Cambridge, Mass.: Harvard University Press, 1983), 41.

60. Wilson, *Human Nature* (Cambridge, Mass.: Harvard University Press, 1978) 133.

61. Edward O. Wilson, *Human Nature*, 119.

62. Edward O. Wilson, *Consilience: The Unity of Knowledge* (New York: Alfred Knopf, 1998), 292.

Chapter 4

Mayr:

Toward a Framework of Mediation

It was the intent of the previous chapter to outline two sharply opposing views of evolutionary biology and its ethical-political implications. Wilson is a leading proponent of a sociobiology which seeks to defend a genetic-cultural coevolution view of qualities pertaining to natural fitness in regard to behavior traits such as gender relations, status, territorial defense, and what he believes to be a genetic altruism in regard to moral evaluation. Gould, Lewontin, Rose, and Kamin, it has been seen, are leading critics of Wilson's sociobiology, which they believe entails an untenable biological reductionism that serves the interest of a conservative political ideology. It is their view that the meaning of Darwinian theory of evolution must be restrictive to simply a concept of biological-cultural *interactionism* without being decisive about what is specifically biological versus what is cultural, and having ethical-political implications that human action and decision entails open-ended possibilities in regard to social change. It will be the intent of this chapter to establish the importance of Ernst Mayr's version of a neo-Darwinian synthesis as a constructive mediation between Wilson versus Gould-Lewontin, Rose, and Kamin. It will be the intent first of all, to show that Mayr's view of Darwinian renewal is congruent with Wilson's emphasis upon the need for consilience of natural science with social sciences and the humanities. But it will be shown that Mayr provides a corrective to the inadequacies of Wilson's view that such consilience must be in the model, a reductive materialism indebted to Bacon and Galileo; Mayr's emphasis upon what is distinctive to biological science.

is the irreducability of higher levels of organic life to lower levels, and his view of a "teleonomic" feature of organic life that is in continuity with Aristotelian naturalism. It will be the intent, second, to show that Mayr's view of Darwinian renewal can be congruent with constructive features in Gould-Lewontin's concept of "punctuated equilibrium," or "local adaptation," while yet believing this contention can be appropriated within a more holistic view of Darwinian natural selection. It will be the intent, third, to show how Mayr's version of Darwinian renewal is roughly convergent with Wilson as a view of the general nature of human evolution, but a mediating position on the question of ethics: a convergence with Wilson as a view of a genetic basis in regard to *capacities*, in ethical adaptation, and in regard to parental and kinship altruism, but closer to Gould, Lewontin, Rose, and Kamin that the main substance of ethics is a product of cultural influence.

But it will be shown that Mayr's approach to ethics is a corrective to Gould-Lewontin, Rose, and Kamin's view of a Darwinian pluralism that does not confront the problems of a substantive content of ethics that Mayr believes necessary to confronting crisis conditions of modernity. It will be finally shown that the contributions of Lawrence Kohlberg, C. H. Waddington, and Julian Huxley, which Mayr invokes as a defense of his view of the ethical implication of human evolution, provides a constructive opening to an ethical-political theory that can be integrative of Darwinian and Aristotelian implications.

I

Mayr's version of a Darwinian renewal entails a significant point of convergence with Wilson's in the emphasis upon the need for a consilience of Enlightenment science with the social sciences and the humanities. Mayr believes that despite the monumental achievement of the seventeenth-century scientific revolution, it created a problem for the Western world in the disjunction of science from the liberal arts. It is biology, he believes, that can provides this bridge. "Evolutionary biology, he contends, shares with history a number of attributes historians have often considered to be diagnostic of history: uniqueness of the treated entities, inability to predict, frequency of tentative (subjective) references, and relevance to religion and morality."[1] But Mayr provides a corrective to the inadequacies in Wilson's view that what we should celebrate as the Enlightenment's achievement is the indebtedness to the reductive materialism of Galileo and Newton. It is there that Wilson's contention seems incoherent. For at one point he does briefly comment that a scientific materialism may be an oversimplification that at each level of living cells and above there are laws and principles that cannot be predicted from those at lower levels. But Wilson does not provide any clarification of what this indicates as an implication of the need for a reconstructive interpretation of the version of Enlightenment science indebted to Galileo and Newton. It is here that Mayr *does* provide such a reconstructive interpretation. Mayr does not object to a *constitutive reductionism* as the "dissection of phenomena, events and processes into constituents of which they are composed."[2] What is more controversial are two types of reductionism. One is that all the phenomena and processes at highest hierarchical levels can be explained in terms of components at lower levels. A second type of reductionism is one that the theories postulated by biology employ case theories and laws formulated in the physical sciences. But Mayr contends that this form of reductionism has been only partially successful even within the physical sciences. "Indeed none of the more complex biological laws has ever been reduced to and explained in terms of the composing single processes."[3] It is Mayr's contention that the attempt to reduce biology to physics has been a failure. Mayr notes that even physical science has been a rejection of strict determinism of classical physics: "For physical laws are statistical in nature and that prediction can be only probabilistic; as well as the development of *concepts* as a tool in understanding of laws governing physical phenomena." Developments in biology, he contends, are thus moving away from the standard concepts of physicalism, essentialism, and determinism.

> Living systems are characterized by a remarkably complex organization which endows them with the capacity to respond to external stimuli, to bind or release energy (metabolism), to grow, to differentiate, and to replicate. Biological systems have the further remarkable property that they are open systems, which maintains a steady state balance in spite of much input and output. This homeostasis is made possible by an elaborate feedback mechanism, unknown in their precision in any inanimate system.[4]

Central to Mayr's reconstructive interpretation is a causality in biology differentiated from classical mechanics, a concept of *teleonomic* processes in living nature as opposed to *teleomatic* processes in inanimate nature. Characteristic of teleomatic processes are physiochemical processes as consequences of natural laws such as the fact that "gravity

provides the end state for a rock I drop into a well."[5] What is distinctive
to *teleonomic* processes of living nature is "goal directed behavior' as
activity associated with food getting and courtship phases of reproduction. A teleonomic behavior is guided by a "program" that depends on
the existence of an endpoint or terminus foreseen in the program that
regulates the behavior.

> The programs which control teleonomic processes in organisms
> are either entirely laid down in the DNA of the genotype (closed
> program) or are constituted in such a way that they can incorpo-
> rate additional information (open program) acquired through
> learning, conditioning or other experiences. Most behavior, par-
> ticularly in higher organisms, is controlled by such a program.[6]

A striking feature of Mayr's view of teleonomic features of organic life
is that he believes it to be a congruency with Aristotelian naturalism.
Mayr recognizes that Aristotle used teleological explanation in geology
and that it was for this reason he was violently rejected by Bacon, Descartes, and other followers. But Mayr believes that Aristotle has been
consistently misunderstood because he used the term *eidos* for his form-
giving principle, and that it was taken for granted that he had something
in mind similar to Plato's concept of *Eidos*. But Mayr believes that
Aristotle had in mind something quite different. He believes it is thus
necessary to rescue the reputation of Aristotle, who has often been called
a "cosmic teleologist." But Mayr believes recent Aristotelian scholars
such as Lennox and Nussbaum are challenging this interpretation: their
contention that Aristotle described the "teleolonomic" processes that
take place in embryonic development.[7] Mayr believes that one achieves a
remarkably modern account in Aristotle's embryological analysis if one
translates his term *eidos* as "genetic program." Max Delbruck, he contends, has pointed out correctly that "Aristotle's principle of the *eidos* as
an 'unmoved mover' is one of the greatest conceptual innovations." In
the view of Delbruck, "unmoved mover" perfectly describes DNA: "it
acts, creates form of development, and is not changed in the process." In
Mayr's view of the existence of teleonomic programs—unmoved mov-
ers—"is one of the most profound differences between the living and the
inanimate world, and it is Aristotle who first postulated such a causation."[8]

II

If Mayr's version of Darwinian renewal can be seen as a framework of constructive dialogue with Wilson's sociobiology in regard to defining the nature of biological science, a similar constructive dialogue can be seen in relation to Gould-Lewontin's concept of a Darwinian pluralism. It was Gould and Miles Eldredge who have been commonly cited as the originators of the concept of "punctuated equilibrium" as an alternative to the orthodox concept of Darwinian natural selection. This involves the central contention that the coherence of a species is not maintained by interaction among its members, but emerges rather as a historical consequence of a species origin in "peripherally isolated populations that acquired its own powerful homeostatic system." What this involves, then, is a challenge to a "synthetic" view of evolution, and the "playing down" of the role of natural selection.[9] Mayr notes that Gould and Lewontin have vigorously attacked the adaptationist program in ways that he is fully in accord with.

Mayr agrees with Gould and Lewontin's attack upon unsupported "adaptionist explanation." They rightly point out the same traits as, for instance, the "grill arches of mammalian embryos had been acquired as adaptations of remote ancestors, but even though they no longer serve their original functions, they are not eliminated because they have become an integral component of a developmental system. Most so-called vestigial organs are in this category."[10] Mayr acknowledges the validity of Lewontin's view that new gene pools are generated in every generation, and that evolution takes place because the successful individuals produced by these gene pools give rise to the next generation. Evolution thus is merely contingent on certain processes articulated by Darwin. Variation and selection is no longer a fixed object transformed, as in transformational evolution, but an entirely new start is made, so to speak, in every generation. "Evolution is no longer necessarily progressive. It no longer strives toward perfection, or any other goal. It is opportunistic, hence unpredictable."[11]

What Darwin did not fully recognize, Mayr points out, is that variational evolution takes place on two hierarchical levels: the level of the deme (population) and the level of the species. Variational evolution at the level of the deme involves "individual selection and leads minimally to the maintenance of fitness of the population through establishing selection." But the second level of variation is that of the species, where there is a more opportunistic production of new species. While most of them are doomed to rapid extinction, a few make revolutionary innovations that give species an improved competitive potential and a "starting point of successful new phyletic lineages and adaptive radiation." Mayr thus acknowledged that the change from transformational to variational evolution was not effectively carried through by most Darwinists. He thus gives credit to Eldredge and Gould in their theory of punctuated equilibrium. "Whether one accepts this theory, rejects it, or greatly modifies it, there can be no doubt that it had a major impact on paleontology and evolutionary biology."[12] Mayr contends that he, himself, was the first to develop a connection between specialization, evolutionary rates, and macroevolution, and that this hypothesis must be the centerpiece of punctuated equilibrium theory. The novelty of this theory is that most rapid evolutionary change does not occur in widespread populous species, but in small "founder populations." This was the conclusion he reached in his study of the speciation of island birds in New Guinea and the Pacific. "My conclusion was that a drastic reorganization of the gene pool is far more easily accomplished in a small founder population than in any other kind of population. . . . Indeed I was unable to find any evidence whatsoever for the occurrence of drastic evolutionary acceleration and genetic reconstruction in widespread populous species."[13]

What is unique to organisms at the molecular level, Mayr contends, is a mechanism for storage of acquired information that inanimate matter does not have. All organisms possess a historically evolved genetic program coded in the DNA of the nucleus which has no counterpart in the inanimate world except in man-made machines. This program consists of a genotype and a phenotype. The genotype is handed down from generation to generation and in interaction with the environment controls the production of the phenotype, which is the visible organism we encounter and study. The genetic program, as Mayr points out, is the process of history that goes back to the origin of life, the "experience of all ancestors." It is this that makes an organism a historical phenomenon, "The genotype also endows such organisms with the capacity for goal directed or teleonomic processes and activities, capacities totally absent in the inanimate world."[14]

It is Mayr's contention that the philosophy of science needs to adopt a greatly enlarged vocabulary that can include such words as "biopopulation," "telenomy," and "program." It will need to abandon past loyalty to a rigid "essentialism" in favor of a broader recognition of "stochastic" process. This will require a greater emphasis upon a pluralism of cause and effect, a greater recognition of a hierarchical organization in much of nature, the emergence of unanticipated properties at higher hierarchical levels, and the "internal coherence of complex system," and many other concepts that have been neglected in the classical philosophy of science.

Mayr believes that "twenty-nine years ago the physicist C. D. Snow described the unbridgeable gap between the physical sciences and the humanities. If biologists, physicists, and philosophers working together can construct a broad-based, unified science that incorporates both the living and the nonliving world, we will have a better base from which to build bridges to the humanities, and some hope of reducing this unfortunate rift in our culture. Paradoxical as it may seem, recognizing the autonomy of biology is the first step toward such unification and reconciliation."[15]

But if Mayr is willing to acknowledge a concept of "local adaptation," he does not believe this is incompatible with the general conceptual framework of Darwinism. Mayr grants that almost any change in the course of evolution might have resulted by chance, although one can never *prove* that this is the case. One can, by contrast, deduce the probability of causation by selection, by showing that respective features would be favored by selection. This is the key to the evolutionary approach. But Mayr emphasizes that Darwin was not defending a "perfectionist" concept of natural selection which is the main brunt of Gould-Lewontin's critique that "is, indeed, easy to ridicule." What Mayr is contending is the need for an "adaptationist" theory that steers a "perilous course between a radical reductionist-atomism and a radical holism."[16] A partially holistic approach is one that asks appropriate questions about integrated components of the system that would not need to be either "stultifying nor agnostic." Such an approach, he believes, may be able to avoid the "Scylla and Charybdis of an extreme atomistic or an extreme holistic approach."[17]

III

Mayr's "Darwinian synthesis" can be seen as mediator between Wilson versus Gould, Lewontin, Rose, and Kamin in his view of where humans fit into evolution. Mayr's view of human evolution follows in general accord with Wilson's in the recognition that what is unique in human characteristics evolved from those of our ancestral animal kingdom. While pre-Darwinian literature saw an inevitable culmination toward ever-greater perfection, biological science embodied the emphasis upon gradual evolution through the process of natural selection. This development also led to the elimination of "essentialism." Darwin's concept of "population thinking" stressed the uniqueness of each individual within the population that also applied to human beings. What made humans possible was the emergence of bipedalism, but more importantly the use of tools and brain enlargement. Mayr believes that selection pressures favoring the evolution of the human brain included the development of speech and the emergence of culture that speech allowed. What was of key importance in human evolution, Mayr believes, was the "social integration in the hominid group." Where the "group as such" became the target of selection, this development facilitated the survival, prosperity, and reproductive success of the group as a whole. This included continuous sexual receptivity, development of the menopause, expanded life expectancy, and other characteristics of humans not found in animals. Among some animals, he believes, the benefits of cooperation are offset to some degree by a potential for conflict, particularly in male competition for females. Such conflict, however, was "mitigated by the cultural trend toward monogamy and social gratification." Incest rules were developed and enforced in most societies. The family became the foundation of the group structure, along with division of labor with men as hunters and women as gatherers. The two sexes thus formed a cooperative unit along with the extended family of grandparents, cousins, uncles, and aunts. It was the extended family that was important not only for mutual help but also cultural cohesion and transmission to the next generation. It is the breakdown of the extended family, Mayr believes, that can be seen as a basic root of cultural breakdown in inner-city slums. Mayr believes that the development of division of labor and job specialization contributed to social stratification in which feudalism was the extreme example.[18]

Where Mayr's view of human evolution is congruent with Gould and Lewontin is in his emphasis upon "punctuation" in the transition from hunter-gatherer to agriculture and animal breeding. Where he is also in congruence with Gould-Lewontin, Rose, and Kamin is in his critique of the role of race in human evolution. Molecular research, he points out, reveals that so-called human races are closely related. This is not to deny differences such as skin color, eye color, hair, shape of nose and lips, skull, and stature. "But when it comes to the psychological characteristics that really count, the role of genes is largely undetermined." Mayr is also strongly critical of so-called "eugenics" as a concept of artificial selection for superior genotypes as a way of lifting humans toward greater perfection. Such a contention, he notes, led to some of the most "heinous crimes of mankind," including the racism that led to Hitler's horrors. Mayr believes that the concept of eugenic measures to bring about "improvement" in the human race is impossible for several reasons. One is the fact that we have no knowledge of the genetic basis for the characteristics we would wish to change. Second, such a possibility would require a mixture of different types of genotypes and nobody has any certainty about what this mixture should be. "Finally, and more importantly, the steps that would have to be taken in order to implement eugenics are simply intolerable in a democracy."[19] In Mayr's view it is precisely the diversity of variation in every human population that forms the basis for a healthy society. This permits a division of labor, but it "also requires a social system that makes it possible for each person to find the particular niche in society for which he or she is best suited." This requires an emphasis upon an ideal of equality that requires equal status before the law, and equal opportunity. This does not mean total identity, for equality is a social/ethical concept and not a biological one. "Neglect of human biological diversity can do harm and can be an impediment in education, medicine, and other endeavors." Mayr believes J. B. Hadane well sums up the political implication of this contention: "It is generally admitted that liberty demands equality of opportunity. It is not equally realized that it also demands a variety of opportunities and a tolerance of those who fail to conform to standards which may be culturally desirable but are not essential for the functioning of society."[20]

It is in regard to Mayr's view of how evolution can account for ethics that his mediation between Wilson versus Gould-Lewontin, Rose, and Kamin becomes most apparent. Mayr is in accord with Wilson on a genetic basis in regard to *capacities* for moral evolution. Mayr cites the view of F. J. Ayala, who affirms that this entails the ability to anticipate the consequences of one's own action, to make value judgments, and to choose alternative courses of action.[21] But, it should be noted, if Mayr is conceding a cultural source of ethics, the *capacity for adopting* ethical behavior leads to the ability to adopt a *second* set of ethical norms "supplementing and, in part, replacing the biologically inherited norms based on inclusive fitness." The basic importance of such norms is that they counteract the basic selfish tendencies of individuals, giving rise to an altruism that benefits the group as a whole, as well as the individual.[22] How individuals acquire such ethical norms, he believes, has been clarified by developmental psychologists such as Conrad Waddington and Lawrence Kohlberg. Their findings support the view that the value system of an individual is incorporated in childhood growth and development that is expressive of the meaning of human evolution as an "open behavioral program" that makes ethics possible.[23] Mayr would argue with Wilson that there are some components of "inclusive fitness altruism," such as in parental care, the moral stance we have toward strangers, as opposed to members of our own group.[24] But Mayr would be more in agreement with Gould, Lewontin, Rose, and Kamin in his contention that the main substance of ethics is the product of culture and human learning experience. The evidence for this, he believes, is indicated in the striking difference in kinds of morality in different ethnic groups; the total breakdown of morality under certain political regimes; the ruthless behavior often displayed against minorities; the often uninhibited bombing of political centers; the warping of character of a child when deprived of a mother during a critical phase of its infancy.[25]

Mayr then raises the question of what kind of ethics will be required to confront crisis conditions of modernity such as the breakdown of the family, drug abuse, and the waste and depletion of our resources. Mayr is convinced that the traditional ethical norms of Western culture are based on those of the Judeo-Christian tradition, the various commandments and injunctions of the Old and New Testaments. But such norms, such as "Thou shalt not kill," are not sufficiently flexible to deal with problems such as abortion or terminal illness. The very core of human ethics must be able to confront conflicting factors in order to make the right choice.[26] A second inadequacy in the ethical norms of the past is that they were addressed to a territorial, pastoral people who did not confront problems of modernized urban society. What this entails is a problem of an "expanding circle" that requires greater effort to overcome past differentiation between ethics employed toward outsiders as opposed to members of one's own group.[27] A further great ethical problem of our time is the excessive egocentrism and attention to rights of individuals that is not accompanied by a consciousness of obligation; the problem of dilemmas that arise when a choice must be made between individual ethics and social or community ethics.[28] A final great ethical problem is that of our responsibility to nature as a whole, and what in recent times has become the conservation environmental ethics articulated by Alto Leopold, Rachel Carson, Paul Ehrlich, and Garret Harden.[29] This will require the development of a "steady state economy, and a readjustment of ethical norms in face of shift from a pastoral agricultural society to the modern industrial world: 'the ethical norms of the future must be flexible enough to evolve as these problems appear if we are to remain an adaptive species."[30] Mayr poses the question of whether there is a particular ethics an evolutionist should adopt. His own values, he concludes, are close to Julian Huxley's evolutionary humanism. "It is the belief in mankind, a feeling of solidarity with mankind and a loyalty toward mankind. Man is the result of millions of years of evolution and our most basic ethical principle should be to do everything toward enhancing the future of mankind. All other ethical norms can be derived from this baseline."[31]

IV

It is finally important to clarify how Mayr's view of ethical implications of evolution are congruent with Gould-Lewontin's emphasis upon the cultural basis of ethics, can be corrective to the inadequacies of a cultural relativism that seem inescapable in their concept of a Darwinian pluralism. It was seen above that it is Mayr's contention that if there can be a genetic causation in regard to the capacities for ethical evolution (the ability to anticipate consequences, make value judgments, choose alternative courses of action,) the main substance of ethics is a produce of learning experience. It was noted that Mayr refers to Kohlberg and Waddington as examples of this approach to ethics, and that he concludes with a brief comment that his own view on an ethics which an evolutionist can support is close to that of Julian Huxley. But it is essential to expand further how these writers provide a reinforcement of Mayr's view of what is required of an ethics that can confront crisis conditions of modernity and how their contribution is a corrective to what is not clarified in the Gould-Lewontin, Rose, and Kamin conclusion that we need nothing more than the last Marxist thesis on Feurbach, namely, that "philosophers thus far have only interpreted the world; the point, however, is to change it."

Central to Lawrence Kohlberg's contribution is what he believes to be a cognitive basis for moral consciousness that emerges from a process of development that is neither biological maturation nor direct learning process, but the reorganization of psychological structures resulting from organism-environment interaction. Cognitive development is a dialogue between the child's cognitive structures and the structures of the environment. It is not an unfolding of instinctual, emotional, or sensorimotor patterns but cognitive changes in general patterns of thinking about self and world, involving thought and symbolic interaction.[32]

A cognitive-developmental theory of moralization, Kohlberg contends, holds that there is a sequence of moral stages for the same basic reason that there are cognitive or logicomathematical stages. This theory would be indebted to Piaget's theory of development, which is closely linked to a normative logic; the notion that stages are forms of equilibrium, forms of integrating discrepancies or conflicts between the child's schemata of action and the actions of others. Kohlberg believes that the sociological concept of "role-taking" is central in this process—the tendency to react to others as like the self's behavior from the other's point of view. The centrality of role-taking for moral judgment is based on the notion of sympathy for others, as well as on the notion that moral judgment must adopt the perspective of the "impartial spectator" or "generalized other," a concept central to moral philosophy from Adam Smith to Robert Firth. Moral principles are cognitive structural forms of role-taking, centrally organized around justice as equality and reciprocity. To role-take in moral situations is the experience of moral conflict; a conflict, for example, between my wishes and claims and the claims of you and a third party. The integration is provided by principles of justice of a stage. The social environment thus stimulates development, providing opportunity for role-taking or for experience of social-moral conflict that may be integrated by justice forms at or above the child's own level.[33]

Kohlberg believes that human moral development can be understood in terms of sequential and hierarchical levels, which he classifies as preconventional, conventional, and postconventional with each level having two discernible stages. The preconventional level includes stage one and two. At stage one what is right is to avoid breaking rules that are backed by punishment, the obedience to superior authority. The social perspective is an egocentric point of view that does not consider the interests of others and where actions are considered physically rather than in terms of the psychological interests of others. Stage two is that of individualism, following rules only when it is in someone's immediate interest and letting others do the same. What is right is also what is fair, an equal exchange, a deal or agreement. The conventional level involves stages three and four. At stage three what is right is living up to what is expected of people close to you in your role as son, brother, friend, etc., and in keeping mutual relationships of trust, loyalty, respect, and gratitude. At stage four what is right involves fulfilling duties to which you have agreed: laws are to be upheld except where they conflict with other fixed social duties. What is right is also contributing to society, the group, or institution. The postconventional level involves stages five and six. At stage five what is right is seen in terms of values and rules relative to your group which are to be upheld in the interest of impartiality and because they are the social contract. There is also the concern that laws and duties be based on a rational calculation of overall utility and the "greatest good of the greatest number." At stage six what is right involves following self-chosen ethical principles such as justice, equality, and human rights, the respect for dignity of individual persons.[34]

While Kohlberg emphasizes moral growth and development as a

product of biological-cultural interaction, he does not specifically address the question of the implication of his contention in regard to Darwinian evolution. It is Waddington who more directly carries out the implications of Darwin's view of the ethical capacities that are products of the acquisition of language, and social habituation in human evolution. For animal species, the mechanisms of human evolution, Waddington contends, have a biological basis in Darwin, but what is essential in ethical development is the learning process and the interaction of the child with his external environment through role-playing, language, and use of symbols.[35] Waddington believes this entails an organismic account of evolution centering upon mutuality and interdependence, rather than what is mechanistic or atomistic. The human species, then, differs from the rest of the animal world. It is Waddington's contention that living things have an evolutionary direction giving rise to a status similar to "healthy growth" and development. Ethical beliefs of man can thus be judged according to their efficacy in furthering an evolutionary direction and in coming to terms with crisis situations in regard to revolution, wars, mass technology.[36] Progress in the world, he believed, entails a direction which can be defended as "psychological health" or as "biological wisdom."[37]

It was seen that Mayr believes his view of the ethical implication of evolution to be close to that of Julian Huxley: that our most basic ethical principle should be to do everything toward enhancing the future of mankind, and that all other ethical norms can be derived from this baseline. What is significant in Huxley's view of evolution can be seen to be congruent with a central point made in chapter 1 that the central ethical implication of Darwin's theory of evolution is not the Spencerian concept of competition and survival of the fittest that is derivative from Darwin's view of animal species, but rather what he saw to be distinctive features of human evolution as the consequence of development of capacities for language, sociality, and habituation. The central contention of Julian Huxley, which he believes to be a corrective to his grandfather Thomas Huxley's belief, is the contradiction between the ethical and the cosmic process that was the result of the development of speech and conceptual thought. "In the first time tradition and education became continuous and cumulative."[38] What this gave rise to was a radical new possibility. Mechanical and natural selection will operate, but will be of secondary importance in comparison to the social and conscious level. "The slow methods of variation and heredity are outstripped by the speedier processes of acquiring and transmitting experience."[39] The mechanism of evolution is no longer blind and automatic, but becomes

conscious and, where ethics can be injected into the evolutionary process, this opens up a new range of possibility, the consciousness of good
and evil, truth and other values, emotional states of love, reverence and
mystical contemplation. Such a development gives rise to several main
possibilities. One is that social organization will not be for the purpose
of simply preventing change, but encouraging it, going beyond a static
stability. A second possibility is one that can promote individual development and self-realization in which individual ethics will be concerned
for equality of opportunity and human well-being.[40]

In Huxley's view, the Golden Rule cannot be put into practice due to
the imperfection of human nature and the lack of a basis for practical
ethics. But it can be approximated in the extension of love and sympathy, the avoidance of suffering, and the stunting of development, the
rational acceptance of moral principles and the means for their realization.[41] This does not mean that, in an imperfect world, pain and conflict
can be eliminated. Ethical principles may entail the belief that war is
wrong, yet it may at times urge that it is right, where it is believed ruthless suppression of the opponent is a duty. Ethics cannot prevent "cosmic injustice" such as congenital deformity, physical disaster, and death
of loved ones. Man is thus an "heir of the past," and a victim of the
present.

But man is not only the heir of the past and the victim of the present:
He is also the agent through whom evolution may unfold the further
possibilities. Here, it seems, is the solution of our riddle of ethical rela-
tivity: The ultimate guarantee for the correctness of our labels of right-
ness and wrongness are to be sought for among the facts of evolutionary
direction. Here, too, is found the reconciliation of T. H. Huxley's antithesis between the ethical and the cosmic process: For the cosmic
process we now perceive is continued in human affairs. Thus man now
imposes moral principles upon ever-widening areas of the cosmic process, in whose further slow unfolding is now the protagonist. He can
inject his ethics into the heart of evolution.[42]

What is significant in Mayr's view of ethical implication of Darwinian evolution and the reinforcement he believes is provided by Kohlberg,
Waddington, and Huxley is what can be fully in accord with Gould-
Lewontin, Rose, and Kamin's critique of the distortions of biological
reductionism that has been of service to conservative political ideologies. As Michael Ruse notes, the Gould-Lewontin theory of Darwinian
pluralism, or "punctuated equilibrium," can be seen as having a significant relation to their left-wing political commitment: Gould's admission,
in fact, that he has praised punctuated-equilibrium theory because it fits

a Marxist "dialectical world picture."[43] What Mayr's version of Darwinian renewal provides is a view of the substance of ethics that can be a corrective to a Darwinian pluralism as simply "open-ended, adaptive stories" that does not formulate criteria for left-wing values commitment that Gould-Lewontin, Rose, and Kamin are obviously endorsing. Mayr's framework is congruent with Gould-Lewontin, Rose, and Kamin's emphasis upon biological-cultural *interactionism*. But what Mayr is contending, as seen above, is that this interaction is the source of a learning experience that can lead to criteria for ethics essential to confronting crisis conditions of modernity. This springs from his contention that if the individual acquires ethical *capacity* that is innate, he then is able to develop a *second* set of ethical norms that are a product of culture. Such ethical norms constitute the potential for restraining selfish, egocentric tendencies in the direction of benefit for the common good of the human community.

What is finally of key significance in Mayr's framework for a Darwinian renewal is what it provides as an opening to integration with Aristotle's ethical-political implication. It was seen that Mayr believes that the program that controls teleonomic processes in organisms is convergent with Aristotle's naturalism. But what is also central to Aristotle's ethics is that if human virtues have a basis in biological *potentialities* of human nature, they must be developed by education and habituation. It will be the intent of the concluding chapter to provide clarification on controversial issues in the question of how it is possible to establish a continuity of Darwinian evolution with Aristotelian implication.

Notes

1. Ernst Mayr, "Biology, Pragmatism and Liberal Education," in *Education and Democracy: Reintegration of Liberal Learning in America*, edited by Robert Urrill (New York: College Board, 1997), 294.

2. Ernst Mayr, *Toward a New Philosophy of Biology: Observations of an Evolutionist* (Cambridge, Mass.: Harvard University Press, 1988), 10.

3. Mayr, *Toward a New Philosophy of Biology*, 11.

4. Mayr, *Toward a New Philosophy of Biology*, 14.

5. Mayr, *Toward a New Philosophy of Biology*, 44.

6. Mayr, *Toward a New Philosophy of Biology*, 49.

7. Mayr, "Biology, Pragmatism and Liberal Education," 292.

8. Mayr, *Toward a New Philosophy of Biology*, 57.

9. N. Eldredge and Steven Gould, "Punctuated Equilibrium: An Alternative to Phyletic Gradualism" in *Models in Paleobiology*, edited by J. M. Schopf (San Francisco: Freeman & Cooper, 1977), 82-115.

10. Mayr, *Toward a New Philosophy of Biology*, 57.

11. Mayr, *Toward a New Philosophy of Biology*, 458.

12. Mayr, *Toward a New Philosophy of Biology*, 459.

13. Mayr, *Toward a New Philosophy of Biology*, 461.

14. Mayr, *Toward a New Philosophy of Biology*, 17.

15. Mayr, *Toward a New Philosophy of Biology*, 21.

16. Mayr, *Toward a New Philosophy of Biology*, 155.

17. Mayr, *Toward a New Philosophy of Biology*, 157.

18. Ernst Mayr, *This Is Biology: The Essence of the Living World* (Cambridge, Mass.: Harvard University Press, 1997), 227-43.

19. Mayr, *This Is Biology*, 246.

20. Mayr, *This Is Biology*, 247.

21. Mayr, *This Is Biology*, 256.

22. Mayr, *Toward a New Philosophy of Biology*, 83.

23. Mayr, *Toward a New Philosophy of Biology*, 84.

24. Mayr, *Toward a New Philosophy of Biology*, 76.

25. Mayr, *Toward a New Philosophy of Biology*, 82.

26. Mayr, *This Is Biology*, 266.

27. Mayr, *This Is Biology*, 267.

28. Mayr, *This Is Biology*, 267.

29. Mayr, *This Is Biology*, 268.

30. Mayr, *This Is Biology*, 269.

31. Mayr, *This Is Biology*, 269.

32. Lawrence Kohlberg, *Essays on Moral Development*, vol. II, *The Psychology of Moral Development* (San Francisco: Harper & Row, 1981), 62-67.

33. Lawrence Kohlberg, *Essays on Moral Development*, vol. I, *The Philosophy of Moral Development* (San Francisco: Harper & Row, 1981), 141.

34. Kohlberg, *The Psychology of Moral Development*, 637.

35. C.W. Waddington, *The Ethical Animal* (Chicago: University of Chicago Press, 1960), 146-48.

36. Waddington, *The Ethical Animal*, 19.

37. Waddington, *The Ethical Animal*, 59.

38. Julian Huxley, "Evolutionary Ethics" in *Darwin*, edited by Philip Appleman (New York: W. W. Norton, 1979), 330.

39. Huxley, "Evolutionary Ethics," 330.

40. Huxley, "Evolutionary Ethics," 333.

41. Huxley, "Evolutionary Ethics," 333.

42. Huxley, "Evolutionary Ethics," 334.

43. Michael Ruse, *The Darwinian Paradigm* (New York: Routledge, 1993), 129-32.

Chapter 5

Aristotle and Darwin

It was the intent of the previous chapter to show that Ernst Mayr's version of a Darwinian synthesis entails a view of goal directed teleonomic character of organic life that he believes is a significant continuity with Aristotelian naturalism. It was also emphasized that Mayr's version of Darwinian renewal is a mediation between Wilson versus Gould-Lewontin, Rose, and Kamin: Mayr is able to agree with Wilson only on conceding the possibility of an "inclusive fitness" altruism as it might apply to parental nurturing, or the different stance we take toward strangers as opposed to members of our own group. There is also, he believes, an inherited *capacity* for ethical evaluation as ability to anticipate consequences, make value judgments, and choose alternatives. But the main substance of ethics is the product of learning experience. In his emphasis upon the role of learning experience, and culturally inherited norms, Mayr is thus convergent with the contention of Gould-Lewontin, Rose, and Kamin. But it was seen in chapter 3 that the latter writers, although obviously committed to left-wing ideals of equality and social justice theory, do not clarify how their version of a Darwinian pluralism is supportive of such ideals. For it was seen that their emphasis is simply upon biological-cultural interactionism; that it is the basis for open-ended "adaptive stories" in which anything seems possible. They do not, then, give any indication that they believe it possible to clarify criteria of adjudication as to the relative merits of rival stories in regard to ethical-political implication. But it was seen in the previous chapter that it is Mayr's view that, where the individual has the *capacity for adopting* ethical behavior, he is able to adopt a second set of ethical norms influenced by culture that enables the counteracting of selfish tendencies toward an altruism that benefits the group as a whole, and which he believes to be the meaning of an evolutionary humanism. It is here that Mayr's contention, although not what he clarifies, provides the key to a constructive synthesis of Darwinian and Aristotelian implications. For it is a central contention of Aristotle's ethical-political theory that if human virtues must be seen as a product of learning and habituation, such virtues have a basis in *potentialities* of human nature. But it is here that one must confront a powerful objection mounted by John Wallach. It is his view that the current forms of Aristotelian renewal are a "curious development" in view of how Aristotle's metaphysical naturalism was a resistance to scientific advances of modernity and how it was applied to a wide range of social, racial, and sexual prejudices. What the current Aristotelian renewal entails, he believes, is a "de-politicizing" and "de-historicizing" of Aristotle's writings, disconnecting the form and substance of his political theory from the historical context that constituted its meaning and scope.[1] Wallach's objection to any attempt to defend the contemporary relevance of Aristotelian categories can also be seen as applying to Darwin's theory of human evolution. For if Darwin's theory is expressive of the Enlightenment break from a classical metaphysics, it can be seen (as noted in chapter 1) as a manifestation of what Darwin reveals in his *Descent of Man* as evidence of racist and class prejudices of the Victorian Age that are alien to twentieth-century historical developments. But what Wallach is missing (as it applies to either what can be defined as the continuing relevance of both Aristotelian and Darwinian contribution) is the nature of hermeneutical interpretation given influential articulation by Hans Gadamer. It is the view of Gadamer that it is one of the illusions of the Enlightenment that there can be a knowledge independent of our "historical being in the world," and what this means as our relation to past traditions. The task of hermeneutical interpretation is thus the understanding of the meaning of a text "for us," differentiated from what is strange or alien and from what may be independent of the intentions of the author.[2] Thus the validity of a contention that Aristotelian categories have a contemporary relevance stems from the conviction that *important features* of his metaphysical biology can be sustained within contemporary scientific and philosophical developments, as well as the contention that within our present-day self-understanding, we are able to contend that Aristotle's concept of human biological nature was given a prejudiced application in the historical context of his time, and that its essential meaning was to be given fuller realization in Enlightenment and post-Enlightenment historical transformation that led to the abolition of slavery and gradual progress in equal political rights for women. The same implication of hermeneutical interpretation can be applied to Darwin. It was noted in chapter 1 that the concept of social Darwinism, in its emphasis upon "competitive struggle" and "survival of the fittest" was largely a derivation from Darwin's *Origin of Species* that pertained to animal life. But what this interpretation neglects is the central thesis of Darwin's *Descent of Man*, namely, the emphasis upon human capacities for language, sociality, habituation enabling the possibility for an ethical orientation to a common good beyond egoism and self-interest. This is not to deny the features of *Descent of Man* that are a manifestation of Darwin's Victorian prejudices in regard to racial difference and class bias that could serve the interests of social Darwinist ideology. But it was emphasized that Darwin's Victorian prejudices are clearly in contradiction with the claim of universality of human capacities regardless of difference of class, sex, or race.

It will be the intent of this final chapter to thus argue that it is possible to defend a constructive continuity of Darwinian evolution with Aristotelian naturalism within the framework of hermeneutical interpretation, but with an important modification of Gadamer's contention that this requires a disjunction from Aristotle's biology because of its metaphysical components. For it will be argued that *important features* of Aristotle's philosophical method and scientific theory can be defended within the framework of an internal or pragmatic realism in which a warranted assertion for a human good is the outgrowth of biological-cultural interaction in human growth and learning experiences. It will be the intent, first of all, to clarify features of Aristotle's naturalism that are congruent with Mayr's view of the nature of biological science and his concept of "teleonomic" features of organic life. It will be the intent, second, to show the convergence of Darwin and Aristotle in regard to the biological basis of ethical evolution: its relation to animal life, but what is distinctive in human development in regard to capacities for speech, sociality, and habituation. But it will be argued that it is Aristotle who provides a more complete elaboration of ethical-political implications only roughly indicated by Darwin, himself, and by Mayr's interpretation of Darwinian theory. It will then be argued that the contemporary relevance of Aristotelian-Darwinian formulation can be a corrective to the inadequacies of utilitarian liberalism, but also a corrective to the so-called communitarian alternative, which is an appeal to Aristotelian renewal and entails an emphasis simply upon the virtues given in social practices of historical narratives that is a disjunction from Aristotle's biological naturalism. It will be then argued that although John Rawls's concept of justice is generally seen as a neo-Kantian formulation, it embodies significant features that can be effectively appropriated within an Aristotelian-Darwinian implication. It will finally be the intent to show how an Aristotelian-Darwinian integration can be defended against the objection that it is committing a version of a naturalistic fallacy or a view that moral principles can be deducible from non-moral biological disposition, and the failure to differentiate the objectivist criteria of scientific investigation from emotive-subjectivist features of ethical inquiry.

I

It was seen in the previous chapter that Mayr seeks to provide a defense for biological science as a break from the reductionism of physical science: his view that higher levels of organic life cannot be reduced to lower levels. Organic life, he believes, is characterized by a "teleonomic" feature that he believes to be congruent with Aristotelian naturalism: a purposive, goal-directed behavior as opposed to a "teleomatic," mechanistic determinism. Organisms have a genetic program coded in DNA of the genotype (a closed program) but also incorporating environmental and cultural influences (an open program). Mayr's conviction that a Darwinian concept of teleonomic features of organic life can be congruent with Aristotelian naturalism is, of course, subject to the obvious objection that Darwinian theory of evolution is a radical break from Aristotle's metaphysical biology. It is thus necessary to show how Mayr is entitled to claim that Aristotle can be rescued from this contention. It is the collaborative effort of Martha Nussbaum and Hilary Putnam that provides an effective framework for defending this contention. It is Nussbaum's contention that Aristotle's naturalism has been misunderstood as a no longer credible metaphysical realism. For central to his method of philosophical inquiry is an emphasis upon the "appearances" that set forth phenomena, as opposed to Platonic essentialism. What this entails is common beliefs and judgments based on our experience of the world, and what emerges from looking at practices. Aristotle's theory of perception is a rejection of the Platonic distinction between appearance and reality: a confrontation with what presents us as conflict, confusion, and contradictions, bringing conflict to the surface, marshalling consideration for and against; a process analogous to that of a competent judge. Aristotle's position, she believes, can be considered a realism that is beyond relativism, and a full-bodied notion of objectivity, but hospitable to a view that "truth is one for all thinking, language-using beings."[3] It is

Nussbaum's contention, then, that the central features of Aristotle's biological naturalism can be defended within an internal or pragmatic realism that she believes has been effectively articulated by Putnam. Here it would be important to emphasize where Putnam differs but also collaborates with Nussbaum's view of Aristotelian naturalism. Putnam agrees that a good deal of Aristotle can be "read in a less metaphysical way than scholars have read into him." But he does not believe that all of his writings can be so read. Putnam comments that:

> The greatest difficulty facing someone who wishes to hold an Aristotelian view is that the central intuition behind that view, that is the intuition that a natural kind has a single determinate form (or nature or essence) has become problematical. . . . The Aristotelian insight that objects have structure is right, provided that we remember that what counts as the structure of something is relative to the ways in which we interact with it. Intentionality and the structure of the world and the structure of language are all intimately related, but it seems that the hope of relating the notion of intentionality to the metaphysical notion of structure (or forms) which itself has no intentional presuppositions, is illusory.[4]

Putnam is a leading exponent of an "internal realism" which would be in accord with hermeneutical interpretation that human understanding cannot lay claim to an a-historical or objectivist view of the universe independent of our conceptual schemes. But Putnam does not believe an internal realism requires resignation to a cultural relativism in which truth claims are simply the "right assertability of one's cultural peers." What we assert to be right is always within a background tradition, but what is given within a tradition presupposes a criteria of reason by which traditions can be criticized. We cannot escape the fact of our pluralism or fallibilism. But one does not have to believe in unique "*best* moral version, or unique *best* causal version; or unique *best* mathematical version; what we have are *better and worse* version, and this is objectivity."[5]

What is significant in Putnam's account of an internal realism is that although he believes it to be in opposition to Aristotle's metaphysics, he nonetheless concedes that what he is defending can be appropriately characterized as an "Aristotelian realism without Aristotle's metaphysics; a defense of a common sense world against the excesses of metaphysics; a middle way between metaphysics and relativism."[6] What

Putnam is clearly implying in this statement is that *important features* of Aristotle's naturalism can be defended within an internal realism. It is this implication that makes possible his collaboration with Nussbaum in defending Aristotle's naturalistic biology. What is central in their collaboration is an effort to show how Aristotle's philosophy of mind is a corrective to both the inadequacies of mind-body dualism as well as a reductive materialism. Aristotle's general analysis, they point out, starts with a general interest in relationships of many kinds in regard to their structure and material organization, including living beings (plants, animals, and humans) as well as nonliving natural beings. It is concerned with changes we see taking place in the world, and their primary "substrata," as well as what accounts give us the "best stories about the identity of things as they persist through time."[7] What is central in Aristotle is the contention that "forms are embedded in ever-changing matter." He differs from most writers on the mind-body problem in that he does not give awareness of the mental any special place in his defense of form. When he argues against materialist reductionism, he is not relying on a "primitive" notion of intentionality. "For the defense of form applies to all substances whether or not they have a mind."[8] But what is central to Aristotle's contention is that physiological accounts cannot provide a causal explanation of animal action; action cannot be explained from the bottom up. "In other worlds there is in perception a transition from potential to actual awareness that is not the transition of any materials from one state to another. There is a psychological transition without material transition. Becoming aware is neither correlated with nor realized in the transition of matter."[9] In his emphasis upon the role of perception, they point out, Aristotle reveals a complex interaction between perceiving and desiring that results in animal movement; where different forms of cognition interact with different forms of desire in order to produce the resulting action. "An animal moves as it does because of the fact that its psychological processes are realized in physiological transition that set up movements that culminate in full fledged local movements."[10] Important functions shared by animals and other creatures are shared by the "soul and body, by perception and memory, and emotion and appetite, pleasure and pain. Their soul and body are *active together* as one thing."[11] Nowhere does Aristotle make a sharp distinction between the cognitive and emotional: for Aristotle perception and desire are activities of the soul realized in some suitable matter. Aristotle's naturalism is thus congruent with Wittgenstein's "preserving the non-reducibility and also the experienced complexity of an internal phenomena such as perceiving, belief and desire," criticizing both naturalist reductionism and Platonic intellectualism. We can have, they concluded, "the nonreductionism and the explanatory priority of the intentional without losing that sense of the natural organic unity of the intentional with its constitutive matter that is one of the great contributions of Aristotle's realism. We suggest that Aristotle's thought really is, properly understood, the fulfillment of Wittgenstein's desire to have a natural history of man."[12]

II

What is indicated above is how the collaborative essay of Putnam and Nussbaum provides a framework for a view of Aristotelian naturalism that is a reinforcement of what Mayr believes is the congruency with a Darwinian theory of biological evolution. But what has been so far considered pertains to general issues of perception and cognition that does not clarify ethical-political implications that can also be seen as integral connections of Darwinian and Aristotelian naturalism. An important feature of this connection needs to be seen as what both Aristotle and Darwin share as the recognition that what is distinctive in human evolution cannot be disjoined from human relationship to animal life. According to Aristotle, it is within the *polis* that human nature is truly expressed and his view is that man is by nature political to *a greater degree* than bees or other gregarious animals (Politics:1223a-15). But Aristotle is nonetheless contending that there is a continuity of human with animal life. As James Lennox points out, for Aristotle human beings are said to begin in the same state as other animals, a similar capability for voluntary action, although lacking excellence of character characteristic of human capacity for *reasoned* choices and *appropriate emotions.*[13] Lennox notes that in his *Parts of Animals* Aristotle comments that we find traits in animals that are similar to humans, just as we see in children the "traces and seeds" of states that will be present in later life; traits that are similar and some analogies. "What is missing in children and other animals is that *integration* of practical intelligence with disposition to feel and act, such that one's feelings and actions tend to be the *appropriate* expression of the life of a rational and political animal."[14]

Darwin's convergence with Aristotle is apparent in his emphasis in *Descent of Man* upon the similarity of human and animal qualities. For animals, he contends, manifest pleasure, pain, happiness, and misery. Animals show the human power of attention as when a "cat watches by a hole, and prepares to spring on its prey." Animals have *memories* for

persons and places. Animals possess some powers of reason as evident
in seeming to pause, deliberate, and resolve. Animals have similar
passions, affection, and emotion and even more complex capacities for
wonder, curiosity, memory, imagination, though in very different de-
grees. Animals, such as chimpanzees, use tools when cracking a native
fruit, somewhat like a walnut with a stone. A rudimentary language is
apparent in gestures of monkeys, and in various types of barking among
dogs that is the expression of feeling.[15]

Darwin recognizes that what is of key importance is the moral sensibility that differentiates human from animal life. But it was noted this
does not mean a radical disjunction from animal life. For it is Darwin's
contention that any animal endowed with well-marked social instincts
would eventually acquire a moral sense or conscience, as soon as his
intellectual powers become as well developed as in man.[16] It was noted
in chapter 1 that this gives rise to *sympathy* with society and fellow
beings, the capacity for images of past actions, and the power of language as a basis for expression of moral obligation. Darwin, it was also
noted, places strong emphasis upon habituation. It is, in fact, difficult to
decide whether certain social instincts as sympathy, reason, imitation are
the indirect result of other instincts or the result of long continued habits.[17] It is habituation, Darwin believes, as recognition of man as a social
being. It is the social instinct, he believes, that is the foundation of a
morality beyond egoism, self-interest, directive to the good of the community.[18] But it would be important to emphasize that the difference
between man and the higher animals is one of *degree* and not of *kind*.
For various senses and intuitions which are celebrated among humans
are found in "incipient and even sometimes in well developed condition
even in lower animals."[19]

While Darwin and Aristotle are thus roughly convergent in their view
of the relation of human to animal life, it is Aristotle who carries out
more fully what this entails as a framework for an ethical-political theory
that is only a possible potentiality of Darwin's *Descent of Man*. Here,
again, it would be important to respond to the objection that a view that
Aristotle can be an appropriation of Darwinian evolution is implausible
in view of what the latter represents as a break from Aristotle's metaphysics of soul, form, and essence that underlies his ethical-political
theory. But as previously noted, the case for the contemporary relevance
of Aristotelian naturalism can be defended in the context of Putnam's
view that *features* of Aristotle's naturalism can be defended in the
framework of a pragmatic realism. The keynote of this possibility stems
from what Aristotle indicates in the opening of his *Nicomachean Ethics*:

his view that politics is not an exact science. In discussing subjects, and arguing from evidence. We must be satisfied with a broad outline of the truth; that is, arguing about what is for the most part so from premises which are for the most part true, and being content to draw conclusions that are similarly qualified (N.E. 1094b 1-25).

Aristotle's ethics thus starts from what was indicated above as the features humans share with animal life, and a consideration of the function of human life that seems to be common even to plants. But in seeking what is peculiar to man, we must exclude the life of nutrition and growth. Next in order is the life of perception, but this is common to all animals. There remains, then, the life of the elements that have a rational principle: "The activity of the soul in conformity with excellence" (1098a,8-27). In considering functions of the soul, Aristotle distinguishes between what is in part rational and part irrational. One part of the irrational soul is vegetative—the source of nutrition and growth. But there is another element of the irrational soul that is receptive to reason, urging men in the right direction and encouraging them to take the best course; indicated in the use of admonition, reproof, and encouragement (1102b 28). Aristotle also speaks of two kinds of virtue: intellectual and moral. Intellectual virtue owes its inception and growth to instruction, but moral good is the result of habit. The moral virtues, then, are engendered in us neither by nor contrary to nature; we are constituted by nature to receive them, but their development is due to habit. Thus all those faculties with which nature endows us we first acquire as potentialities, and only later effect their actualization. All the virtues we acquire are the result of exercising them, as in becoming builders by building; instrumentalists by playing instruments. Similarly, we become just by performing just acts, temperate by performing temperate acts; brave by performing brave ones. This view, he believes, is supported by what happens in a city-state. Legislators make their citizens good by habituation (1103b.1-5).

If virtues are a product of learning and habituation, there is little doubt that Aristotle believes this to be directive to the possibility of a general good beyond simply what is given within a particular social practice or historical contingency. Aristotle contends that every art, every investigation aims at some good, as in medical science, health; military science, victory; economic science, wealth. If, then, our activities have some end which we want for its own sake, and for the sake of which we want all other ends, it is clear that it must be the Good that is the supreme good (1097a.1-30).

But a central feature of Aristotle's ethics is the concept of prudence or practical wisdom. Scientific knowledge is demonstrative, the judgments having to do with first principles and universals. Prudential judgment is concerned with the variable and particulars that are distinctive to the sphere of human conduct. This is why people who do not possess theoretical knowledge may not be more effective in action than those who do possess it (1141b 1-20).

What is central in Aristotle's account of virtues, then, is that any account of conduct can be stated only in a broad outline rather than in precise detail, where application to particular problems admits of no precision. What is essential is the clarification of moral qualities that can approximate a "mean" between excess and deficiency; where, for example, in extremes of fear and confidence the mean is courage; in domain of pleasures and pains the mean in temperance; in a field of giving and receiving money the mean is liberality. It is for this reason, Aristotle believes, that it is difficult to be good, because in any case it is difficult to be certain about the mean (1109a 1-15).

Aristotle's *Politics* is a further indication of what can be integrative of Darwinian and Aristotelian categories. Darwin's *Descent of Man* is clearly convergent with Aristotle's view that the state is a creature of nature, but man is by nature more of a "political animal" than bees or other animals due to the gift of speech as a basis for setting forth what is expedient and inexpedient; just and unjust (Politics: 1253a. 15). Aristotle is thus carrying out more fully what is only generally indicative in Darwin's view of human evolution. According to Aristotle, "Every state is a community of some kind and every community is established with a view to some good." It is the state, or political community, he contends, that embraces all the rest, aiming at a good in a greater degree than any other and at the highest good. The integral connection of Aristotle's political theory with his naturalism is clearly evident in his view of the origin of the state. The family, he contends, is the association established by nature for the supply of basic needs. When several families are united, and association aims at more than supplying basic needs, the first society to be formed is the village, and when villages are united into a community large enough to be self-sufficient, the state came into existence, "originating in the bare need of life and continuing for the sake of the good of life" (1253a. 1-10).

It is at this point necessary to confront the common contention that Aristotle's ethical-political theory entails metaphysical realism integral to his biological naturalism: the concept of a *telos* of nature. If Aristotle believes the *polis* is the quest for the highest good, he rejects the concept of a unitary state. For the state is not made up only of so many men, but of different *kinds* of men: for similars do not constitute a state. It is not a military alliance" (1261a.15-25). If Aristotle believes reason must be a restraint upon passion, he is not assuming this is the imposition of an arbitrary authority. For it is rather a government based upon the rule of law that is the best approximation to justice. "Therefore he who bids the law rule may be deemed to bid God and Reason alone rule, but he who bids man rule adds an element of the beast, for desire is a wild beast, and passion perverts the minds of rulers even when they are the best of men. The law is reason unaffected by desire" (1287a.15-30).

The pragmatic features of Aristotle's political theory are particularly evident in his emphasis upon the need for a mixed regime as balance between conflicting claims to power which in some circumstances should lean toward democracy, but in other circumstances lean toward oligarchy. But in all cases a mixed regime would be constructed in accordance with kinds of institutions that local tradition and experience would make most acceptable, and this would entail some mixing of egalitarian and inegalitarian claims to power (1284a. 5-15).

III

In seeking to establish the contemporary relevance of an Aristotelian-Darwinian integration in regard to the ethical-political theory, it is necessary to clarify this possibility as a basis for a reconstructive liberalism that can be a corrective to the inadequacies of classical liberal utilitarianism as well as the so-called communitarian alternative. It was noted in the introduction that the tradition of classical liberalism emerged out of the early Enlightenment break from Aristotelian naturalism toward a mechanistic determinism influenced by Galileo, Newton, and Descartes in which laws realized in nature are not (as Aristotle believed) an expression of what *ought* to be but what *must* be.[20] As George Sabine points out it was this development that carried into the political theory of Hobbes: that which controls human life is not an end, but a cause, of psychological mechanism of the human animal. A state of nature is one of perpetual desire for power and self-preservation. While John Locke had a more benign view of human nature as bases for human rights and duties, his view was also fundamentally egoistic in terms of pleasure and pain. Locke's influence carried over into the tradition of nineteenth-century Benthamite ultilitarianism in which the greatest happiness of the "greatest number" is the measure of good government.[21] Few would wish to deny striking achievements of modernity that brought about the abolishment of slavery, the spread of constitutional government and liberty, and the economic and technological innovations that have enriched the quality of life and well-being in modern industrial societies. Yet the developments of the post-Enlightenment era have been witness to the advent of colonialism; imperialism; totalitarian extremism of left and right; catastrophic wars; the economic structure of capitalism that have enthroned a "calliclean" of endless consumption, growth, and bigness; the persistence of inequalities and social injustice; environmental degradation due to the ascendency of market priorities; the loss of genuine human community and civic virtue.

While there is a widely shared consensus upon distortion resulting from the tradition of utilitarian-individualism, what is more controversial is a direction of reorientation. One influential development has been the neo-Aristotelian communitarian position given most influential articulation by Charles Taylor, Michael Sandel, and Alisdair McIntyre. This entails a general emphasis upon human identity established in individual involvement with tradition and social practices and historical narratives, and upon the concept of Aristotelian *praxis* as a form of contextually embedded, structurally sensitive judgment of particulars.[22] This development provides a constructive appropriation of features of Aristotle's ethical-political theory that were outlined above, but what it lacks is sufficient cognizance of what was emphasized as the integral connection of Aristotle's ethical-political theory with his biological naturalism. What this so-called communitarian interpretation also lacks is a basis for showing that distortions of utilitarian liberalism do not require the conclusion that there cannot be a constructive achievement of the Enlightenment heritage that can be salvaged from the distortion of utilitarianism. John Rawls's neo-Kantian approach to justice has become the most influential articulation of this possibility. Rawls is affirmative of what is central to the contribution of Kant as a view that what is essential to justice must be the human desire to treat men as ends in themselves and never as means.[23] The central deficiency of utilitarianism is its emphasis upon the view that the principle of social justice is an "aggregative conception of the welfare of the group" and where there is no reason in principle why the greater gains of some should not compensate for the lesser losses of others, "or why the liberty of the few might not be made right by the greater good shared by many."[24] Thus classical utilitarianism is in conflict with a Kantian principle that to regard persons as "ends in themselves" in the basic design of society is to agree to forego those gains which do not contribute to everyone's expectation. Utilitarianism, by contrast, by regarding "persons as *favored* 'means' is to be prepared to impose on those already less favored still lower prospects for the sake of higher expectations of others."[25]

Rawls's neo-Kantian theory of justice is that principles of justice for the basic structure of society are the object of an "original agreement" upon the principles among free and rational persons concerned to further their interests in an initial position of equality as defining the fundamental terms of their association.[26] The principles of justice are chosen behind "a veil of ignorance" in regard to one's place in society, class position, social status, natural assets, and abilities. This entails a consensus upon primary goods: basic rights and liberties, powers and prerogatives of office, income and wealth, and the basis of self-respect. Justice also encompasses a "difference principle" in which economic inequalities are allowed so long as this improves everyone's situation, including that of the least advantaged.[27]

Rawls's concept of justice, as outlined above, stems from Kantian categories in which it is his concern to define a "thin theory of the good" that underlies the choice of principles in the "original position." His reason for this, he points out, is that justice as fairness is a concept of "right over the good" that is in contrast to a teleological view of the good or comprehensive moral, religious doctrines on which agreement is not possible in the pluralism of modernity.[28] But Rawls contends that it is nonetheless necessary to provide a deeper understanding of a concept of the good that encompasses moral psychology and the acquisition of sentiments of justice.[29] What this requires is a concept of a deliberative rationality, a rational plan for a person in considering consequences of various courses of action open to him in the realization of his more fundamental desire.[30] It is here that significantly Aristotelian implications of Rawls's theory of justice begin to emerge. In his discussion of the concept of "goodness as rationality," Rawls speaks of an "Aristotelian principle." A person's goal is determined by a rational plan of life that he would choose with deliberative rationality from the maximal class of plans. Such a principle recognizes broad features of human desires and needs, their relative urgency and cycles of recurrence, and other phases of development as affected by physiological and other circumstances.[31] Also a deliberative plan must be consistent with trends in human maturation and growth and how human beings can be best trained and educated for whatever purpose. The Aristotelian principle, then, is that "other things equal, human beings enjoy the exercise of their realized capacities, their innate or trained abilities, and this enjoyment increases the more the capacity is realized, or the greater its complexity."[32] The Aristotelian principle expresses a psychological law governing our desires, a person's capacities that are brought about by physiological and biological maturation.[33] Rawls points out that the Aristotelian principle is borne out by behavior of children and higher animals and seems susceptible to an evolutionary explanation in which "natural selection must have favored creatures of whom this principle is true."[34]

A further Aristotelian implication is indicated in Rawls's view of justice in a "well-ordered society." This is a society where everyone accepts and knows that others accept the same principles of justice and the social institutions this requires.[35] A significant feature of justice in a "well-ordered society," Rawls believes, entails "natural sentiments" that are the product of learning experience in human growth and development.[36] This involves three stages. One is the "morality of authority" that is developed in a child's relationship to the authority of his parents: the love and trust this entails and the sense of worth of his own person.[37] A second stage is the "morality of association" that gives rise to consciousness of duty and obligation. Standards of conduct that are acquired in family, school, and neighborhood; the development of cooperative virtues of justice and fairness, fidelity and trust, integrity and impartiality.[38] A third stage is the "morality of principle" that is central to principles of justice. This leads to an acceptance of just institutions and our part in these arrangements, along with a willingness to work for the setting up of just institutions, or to reform them when justice so requires.[39]

In his consideration of a sense of justice, Rawls thus places strong emphasis upon the connection between "moral and natural attitudes." Such natural sentiments develop from the child's relationship within the family, as well as within associations that give rise to friendship and mutual interest. The principles of moral psychology, Rawls contends, entail three psychological laws. One is the fact that if family institutions express love and caring for his good, the child comes to love them. Second is that if social institutions are just and publicly known, a person will develop ties of friendly feeling and trust toward others in these associations. A third law is that given the presence of conditions of the first two laws, the person will develop a corresponding sense of justice as he "recognizes that he and those for whom he cares are the beneficiaries of these arrangements."[40]

It is significant, then, that in his emphasis upon the possibility that principles of justice are consistent with a view of human evolution as well as his concept of natural sentiment, Rawls's theory would seem to be congruent with a "naturalistic ethics." But Rawls refuses this implication: "While some principles may *seem* natural and obvious, there are great obstacles in maintaining they are necessarily true or even explaining what is meant by this."[41] It should also be emphasized that Rawls believes that principles of justice essential to the structure of constitutional democracy must be characterized as *political* in contrast to more comprehensive moral, philosophical, and religious doctrines on which agreement is not possible in the pluralism of modernity. The concept of justice is not its being true either as having a natural basis or derivative from an antecedent moral order, but since a congruency within our self-understanding within the history and traditions of our public life.[42] What this entails, then, is the possibility of an "overlapping consensus" between rival comprehensive moral, political, or religious views in regard to the fair terms of cooperation between citizens regarded as free and equal and what, as noted above, is the substantive content of a concept of justice in regard to basic rights and liberties, and primary goods.

It is at this point that Rawls is subject to William Galston's charge that his theory of justice is divided against itself: "Explicitly Kantian but implicitly Hegelian in abandoning Kantian standpoint above history and culture," and losing the force of Kant's view that moral principles are not simply a product of social practices.[43] But, it should be noted, Rawls is clearly not contending that principles of justice are authoritative *simply* because of their having the sanction of particular historical traditions or social practices. The process of justification, he contends, is how a moral judgment fits in with and organizes our considered judgments in "reflective equilibrium." Justification is a mutual support of many considerations, of everything fitting together into a coherent view, where first principles and particular judgments appear on the balance to hang together reasonably well in comparison with alternative theories. Justice as fairness is thus a hypothesis that principles chosen in an original position are identical with those that match our considered judgments and so these principles describe our sense of justice. But such judgments must make allowance for possible irregularities and distortions so that when a person is presented with an initially appealing account of his sense of justice, he may well revise his judgment to conform to its principles even though the theory does not fit his existing judgments exactly. From the standpoint of moral philosophy, then, the best account of a person's sense of justice is not the one which fits his judgment prior to his examining any concept of justice, but rather the one that matches his judgment in reflective equilibrium. "This state is one reached after a person has weighed various proposed conceptions and has either revised his judgments to accord with one of them or held fast to his initial conviction."[44]

As seen above, Rawls is skeptical toward any possibility of defending his theory of justice as a naturalistic ethics. But Rawls's concept of reflective equilibrium, it can be contended, is essentially similar to Putnam's concept of an internal or pragmatic realism which (as noted above) enables him to be collaborative with Nussbaum in defending Aristotle's naturalism without believing this requires recourse to a no longer credible metaphysical realism. While human understanding cannot be disjoined from our conceptual schemes, this does not require resignation to a cultural relativism.[45] It is Putnam's view that the fact that we cannot escape the fact of our pluralism and fallibilism does not preclude warranted assertability for better or worse versions, and "this is what we mean by objectivity."[46] Putnam's pragmatic realism can thus be joined with Rawls's concept of reflective equilibrium as a basis for contending that the concept of justice as fairness can be defended as a minimal content of justice having a possibility of cross-cultural consensus and thus supportive of the case for a naturalistic ethic that is integrative of Aristotelian and Darwinian implications.

IV

It is finally necessary to confront the common objection that the concept of a naturalistic political theory entails the so-called naturalistic fallacy in assuming moral ideals can be derivative from nonmoral facts about biological disposition, and the confusion of scientific with moral evolution. In confronting this charge, it is important, at the outset, to emphasize that Mayr's version of a Darwinian synthesis, as considered in the previous chapter, is not defending a view that the substance of ethics can be derived from genetic causation. What can be defended as having a genetic basis is only *capacities* to anticipate consequences, the ability to make value judgments, to choose alternative courses of action. But Mayr is nonetheless committed to the view of Julian Huxley in regard to an ethics that can be an advancement of the evolutionary process as a solidarity with nature as a whole. But it would be important to emphasize that what Huxley recognizes as an ethics that is both a continuity as well as break from biological dispositions of species evolution. It is here, he tells us, that he is in disagreement with his grandfather, Thomas Huxley, who believed that the meaning of Darwinism must be seen as a break from the "cosmic process" of struggle for existence, suffering, and injustice. But it was noted in the previous chapter that it was the view of Julian Huxley that ethical development must recognize the continuing presence of the "cosmic process" in human development, the persistence of chance and its immorality into human life; the fact that man is heir and victim of the past as well as the agent through which evolution may unfold. The problem, then, is how man can impose moral principles upon the cosmic process, and inject ethics into the heart of evolution.[47]

What is significant in Julian Huxley's view of the relationship of ethics to evolution, then, is not an assumption that ethical vision is simply a deduction from *facts* of evolution, but what ethical vision entails is the outcome of the process of reconciling conflicting claims and interests. It was noted in chapter 2 that Mary Midgely clarifies the meaning of this contention. While Midgely does not subscribe to Wilson's concept of a "genetic altruism," she does endorse a Darwinian-Aristotelian integration in which distinctive attributes of man in regard to speech, rationality, and culture are not something opposed to nature, but continuous with and growing out of it.[48] But, again, Midgley is not subscribing to a view of an ethical naturalism that entails the derivation of ideal norms simply as deduction from biological drives and dispositions. What she does believe is that moral ideals do require taking into account facts about human needs and wants. If I say, for example, that playing with children is natural, I am not automatically saying this is laudable or praiseworthy. I am rather noting advantages as well as dangers in forbidding or neglecting it. What we are confronting is a multiplicity of wants and needs in which conflict is at the heart of the problem; where, for example, "love clashes with honor, order with freedom, justice with prudence."[49] What is central, then, in ethical inquiry is the problem of balancing conflicting wants and needs, and it is here Midgely is fully in accord with the features of Aristotle's ethical-political theory (previously noted); the role of practical reason as the cognizance of relations universal to historical particulars, balancing conflicting claims to power, the determination of the mean between excess and deficiency in regard to particular dispositions.

But it is still necessary to confront the objection that if exponents of ethical-political naturalism are not guilty of deducing ethical norms from particular biological dispositions, they are, nonetheless, failing to recognize that the process of critical inquiry must be a clear differentiation between the impartial, objective criteria of scientific method versus the subjective, emotive character of ethical evolution. Such an objection is obviously well warranted as it applies to the disproportion of scientific evidence on behalf of the racist, sexist, class prejudices of so-called social Darwinism considered in chapter 1, and is what Gould-Lewontin, Rose, and Kamin further elaborate on this development as noted in chapter 3. But it was also seen that the strongest criticism of this distortion was mounted by social scientists in the early twentieth century, who were strongly motivated by their moral aversion to the types of biological reductionism that were the basis of denial of civil rights and equality of opportunity to racial minorities and women. It was seen in chapter 3 that Gould-Lewontin, Rose, and Kamin, if concerned with the scientific distortions that have been due to biological reductionism, they are also strongly motivated by their sympathy toward a left-wing, neo-Marxist political orientation. But it was seen that their version of a Darwinian pluralism is simply an emphasis upon biological-cultural interactionism in which the meaning of human evolution is simply open-ended "adaptive stories" in which "anything is possible." They do not, then, seek to show that their commitment to left-wing ideals of equality and social justice is what could be seen as the implication of Darwinian pluralism. It was seen in the previous chapter that Mayr, although rejecting a genetic explanation of the substance of ethics, strongly believes that the concept of Darwinian evolution points to the need for an ethics overcoming the rigidity of egocentrism and directive to consilience with nature as a whole. It was noted that Mayr believes what he is defending is congruent with a concept of learning experience articulated by Waddington in which what is distinctive in human evolution in regard to capacity for language and symbols, role-playing, and culture is the basis for an ethics that can be judged according to efficacy in furthering an evolutionary direction that can be defined as a "psychological health" or "biological wisdom." Such a contention, of course, remains subject to the objection that they are confusing scientific claims with ethical evaluation. What needs clarification, then, is the contention that the fact that scientific evidence has frequently been distorted on behalf of ideological commitments does not require the conclusion that a viable theory of Darwinian evolution requires the disjunction of scientific from ethical inquiry. It is here that Hilary Putnam's view of a pragmatic realism, noted above as a defense of Aristotle's naturalism, is also a basis for challenging the so-called fact/value dichotomy that is a continuing inheritance from the tradition of logical-positivism. Central to Putnam's concept of a pragmatic realism entails the view that we can no longer adhere to a metaphysical realism as the intrinsic property of things apart from the contribution of language and mind. The rejection of a "spectator view" in scientific inquiry, he points out, has given rise to an emphasis upon the role of conceptual schemes in scientific practice: the role of "metaphor" and "world pictures" indispensable to the understanding of our experience.[50] It is here, he believes, that there is a similarity between scientific and moral evaluation. For the acceptability of a scientific theory has to do with whether or not it exhibits certain virtues that are part of our idea of *human cognitive flourishing*; where we are interested in "coherence, comprehensiveness, functional simplicity."[51] This is not to say ethical values are objective in the sense of a laboratory or deductive science, but that "virtues" as coherence or functional simplicity stand for properties of things, that are not just feelings of the persons who use the terms.[52] It is thus valid to speak of a broader criteria of "rational acceptability" in which talk about tables and chairs is comparable to terms we use to describe people as *considerate* or *inconsiderate*, which may be used to praise or blame, but may also be used to explain and predict.[53] What Putnam challenges, then, is that ethical evaluations are simply "projections" or observations that are just "feelings" having no objective properties. But in Putnam's view a sense of justice and an idea of the good, such as what is involved in our reaction to an atrocity is one where we see similarity between injuries to others and injury to ourselves, benefits to others and to ourselves. We thus invent moral worlds for morally relevant features of situations that are more sophisticated than projection theory. Putnam realizes moral perceptions are not reducible to the "world pictures" of physics, but rather that it is possible to defend Aristotle's view that is there is no possibility of a "science of ethics," we can be content with a truth about things, "which are only for the most part true, and with premises of the same kind, we can indicate the truth roughly and in outline.[54] What is central to Aristotle's ethics, as Stephen Salkever notes, is the interconnection of explanation and evaluation. Aristotle's view is thus similar to the explanation we constantly employ when we say a legal system is "repressive," an employer "negligent" in providing for safety of workers. Such explanations are then implicitly evaluative; not only in regard to private tastes but also to a "definitive species-specific context." Salkever also notes how metaphors are central to Aristotle's science, well illustrated in a "medical technology." Just as the point of medical science is to cure particular individuals, so the point of social science is to provide guidance in improving our political life. Thus some "general theoretical grasp of what constitutes health and what constitutes a good *polis* may well be one of the necessary conditions for an adequate pursuit of these goals."[55]

It would be important to note that what Putnam believes is at least some approximation to convergence of ethical and scientific theory in the area of physical sciences, becomes more of a convergence in biological science, which, as Mayr contends, cannot be reducible to the mechanistic laws of the physical sciences. It is for this reason, he believes, that it is necessary to overcome the chasm between science and the liberal arts fostered by the seventeenth-century scientific revolution. Biological science, he believes, provides that bridge, and certain biological disciplines are in fact closer to the humanities than physics. For "evolutionary biology starts with history on a number of attributes historians have considered to be diagnostic of history: uniqueness of treated entities, inability to predict, frequency of tentative (subjective) inferences and relevant to religion and morality."[56] The previous dominance of the Vienna Circle, he contends, was a philosophy of science based on logic, mathematics, and physics. But a philosophy of biology, based on Darwinian thinking, has a greater emphasis upon concepts rather than laws. The importance of concepts, he contends, thus encompasses worldviews such as democracy, freedom, altruism, progress, and responsibility.[57] Mayr's view of Darwinian biological science is thus fully congruent with the meaning of Aristotelian naturalism.

It would be important, finally, to emphasize a key point of convergence between Aristotle and Darwin. What is central to Aristotle's ethical-political theory, as previously noted, is that what is given as *potentiality* of a human biological nature for human virtues must be developed by habituation and education; his view that the polis originates in biological needs and necessities, it continues for the sake of the good life; his recognition that man is more of a political animal than bees or any other gregarious animal endowed with the gift of speech that enables a differentiation of what is expedient or inexpedient, just or unjust.

It was noted in chapter 1 that Bernard Yak believes that writers such as Hannah Arendt have wrongly attributed to Aristotle a dichotomy between politics, freedom, and humanity versus social needs and biological necessity. For Aristotle clearly states that both our distinctiveness as human beings and our attachment to political life grows out of our biological nature. "We are by nature political animals because we are naturally inclined to live in communities and because our natural capacity for speech and argument leads us to form specifically political communities." Yak points out that Aristotle fully recognizes that we need law and political education to live a fully human life. "But the training of virtues is not for Aristotle a fight against nature and certainly not a struggle to transform naturally self regarding beings into other-regarding citizens. It is, instead, a process in which we draw out and build on human beings' natural capacity and natural impulses for communal living."[58]

It was also noted in chapter 1 that Darwin's *Descent of Man* can be seen as a point of significant congruency with Aristotle. This is most apparent in his contention that any animal whatever endowed with marked social instincts would inevitably acquire a moral sense or conscience, as soon as its intellectual powers had become as well developed or nearly as well developed as in man. This entails the pleasure in society of its fellows and sympathy for them; the role of language in expression of opinions on the common good; the centrality of habituation as development of a consciousness of "ought" or duty and obligation.[59] A key Aristotelian implication is also apparent in his contention that the social instincts have been developed for the general good of the community, rather than the school of morals as simply a form of selfishness or the "greatest happiness principle."[60] What needs to be emphasized, in conclusion, is the unity of Darwin and Aristotle in seeking to overcome the disjunction of moral evolution from a human biological nature.[61]

Ernst Mayr's version of the contemporary Darwinian synthesis, as outlined in the previous chapter, provides the key to what can be effectively sustained as the integration of Darwinian and Aristotelian naturalism. While Mayr is not endorsing past forms of biological reductionism, he emphasizes the distinctive features in human evolution from animal species in regard to *capacities* for language and speech that is convergent with Aristotle. Mayr is also convergent with Aristotle in believing that capacities for ethical evaluation include being able to anticipate the consequences of actions, make value judgments, and choose between alternative courses of action. The convergence with Aristotle is further evident in his view that the capacities for ethical evaluation can be appropriately characterized as a "second nature" that is able to counteract excessive egocentricity toward what can benefit the well-being of the group as a whole. Mayr thus provides a persuasive basis for the conjunction of Darwinian and Aristotelian vision as an "ethical system that can maintain a healthy human society and provide for the future of the world preserved by the guardianship of man."

Notes

1. John Wallach, "Contemporary Aristotelianism," *Political Theory* (November 1992): 635.

2. Hans Gadamer, *Truth and Method* (New York: Crossroad Press, 1982), 259.

3. Martha Nussbaum, *The Fragility of Goodness* (New York: Cambridge University Press, 1993), 240-63.

4. Hilary Putnam, "Aristotle after Wittgenstein," in *Words and Life: Hilary Putnam* (Cambridge Mass.: Harvard University Press, 1994), 79.

5. Hilary Putnam, *The Many Faces of Realism* (LaSalle, Ill.: Open Court, 1987), 77.

6. Hilary Putnam, "The Dewey Lectures 1994," *Journal of Philosophy* (September 1994): 447.

7. Hilary Putnam with Martha Nussbaum, "Changing Aristotle's Mind" in *Words and Life*, 24.

8. Putnam, "Changing Aristotle's Mind," 25.

9. Putnam, "Changing Aristotle's Mind," 31.

10. Putnam, "Changing Aristotle's Mind," 35.

11. Putnam, "Changing Aristotle's Mind," 39.

12. Putnam, "Changing Aristotle's Mind," 55.

13. James Lennox, "Aristotle on the Biological Roots of Virtue," in *Biology and the Foundation of Ethics*, edited by Jane Maienschen and Michael Ruse (New York: Cambridge University Press, 1999), 16.

14. Lennox, "Aristotle on the Biological Roots of Virtue," 25.

15. Charles Darwin, *The Descent of Man and Selection in Relation to Sex*, introduction by John Tyler and Robert M. May (Princeton, N.J.: Princeton University Press, 1981), 54.

16. Darwin, *Descent of Man*, 71.

17. Darwin, *Descent of Man*, 72, 77.

18. Darwin, *Descent of Man*, 98.

19. Darwin, *Descent of Man*, 105.

20. George Sabine, *History of Political Theory* (New York: Henry Holt, 1950), 460.

21. Sabine, *History of Political Theory*, 462.

22. Seyla Benhabib, "In the Shadow of Aristotle and Hegel: Communicative Ethics and Current Controversies in Practical Philosophy," *Philosophical Forum Quarterly* (fall-winter 1989-1990): 3.

23. John Rawls, *A Theory of Justice* (Cambridge, Mass.: Harvard University Press, 1999), 156.

24. Rawls, *A Theory of Justice*, 23.

25. Rawls, *A Theory of Justice*, 157.

26. Rawls, *A Theory of Justice*, 10.

27. Rawls, *A Theory of Justice*, 118, 54, 87.

28. Rawls, *A Theory of Justice*, 22; John Rawls, *Political Liberalism* (New York: Columbia University Press, 1993), 36.

29. Rawls, *A Theory of Justice*, 347.

30. Rawls, *A Theory of Justice*, 372.

31. Rawls, *A Theory of Justice*, 372.

32. Rawls, *A Theory of Justice*, 374.

33. Rawls, *A Theory of Justice*, 375.

34. Rawls, *A Theory of Justice*, 378.

35. Rawls, *A Theory of Justice*, 397.

36. Rawls, *A Theory of Justice*, 402.

37. Rawls, *A Theory of Justice*, 405.

38. Rawls, *A Theory of Justice*, 409.

39. Rawls, *A Theory of Justice*, 414.

40. Rawls, *A Theory of Justice*, 425-29.

41. Rawls, *A Theory of Justice*, 506.

42. Rawls, *Political Liberalism*, 13.

43. William Galston, *Liberal Purposes: Goods, Virtues and Diversity in the Liberal State* (New York: Cambridge University Press, 1991), 136.

44. Rawls, *A Theory of Justice*, 43.

45. Putnam, *The Many Faces of Realism*, 70.

46. Putnam, *The Many Faces of Realism*, 77.

47. Julian Huxley, "Evolutionary Ethics," in *Darwin*, edited by Philip Appleman, 334.

48. Mary Midgley, *Beast and Man: The Roots of Human Nature* (New York: Routledge, 1979), 321.

49. Midgley, *Beast and Man*, 186.

50. Putnam, *The Many Face of Realism*, 21.

51. Putnam, *Reason, Truth and History* (New York: Cambridge University Press, 1989), 135.

52. Putnam, *Reason, Truth and History*, 135.

53. Putnam, *Reason, Truth and History*, 138.

54. Putnam, *Reason, Truth and History*, 135.

55. Stephen Salkever, *Finding the Mean: Theory and Practice in Aristotelian Political Philosophy* (Princeton, N.J.: Princeton University Press, 1990), 99.

56. Ernst Mayr, "Biology, Pragmatism and Liberal Education," in *Education and Democracy: Reshaping Liberal Learning in America*, edited by Robert Urrill (New York: College Board, 1997), 294.

57. Mayr, "Biology, Pragmatism and Liberal Education," 290.

58. Bernard Yak, *The Problems of a Political Animal* (Berkeley: University of California Press, 1993), 12, 15.

59. Darwin, *The Descent of Man*, 70-92.

60. Darwin, *The Descent of Man*, 97.

61. Ernst Mayr, *This Is Biology: The Essence of the Living World* (Cambridge: Mass.: Harvard University Press, 1997), 270.

Bibliography

Appleman, Philip. "Darwin among the Moralists." In *Darwin*, selected and edited by Philip Appleman, 551-71. New York: W. W. Norton, 1970.

Aristotle. *The Complete Works of Aristotle*. The Revised Oxford Translation. Edited by Jonathan Barnes. Princeton, N.J.: Princeton University Press, 1984.

Benhabib, Seyla. "In the Shadow of Aristotle and Hegel: Communicative Ethics and Current Controversies in Practical Philosophy." *Philosophical Forum Quarterly* (fall-winter 1989-1990).

Bernstein, Richard. *The New Constellation: Ethical-Political Horizons of Modernity*. Cambridge, Mass.: M.I.T. Press, 1992.

———. *Philosophical Profiles*. Philadelphia: University of Pennsylvania, 1986.

Carnegie, Andrew. *The Gospel of Wealth in Darwin*. Edited by Philip Appleman. New York: W. W. Norton, 1979.

Darwin, Charles. *The Descent of Man and Selection in Relation to Sex*. Introduction by John Tyler and Robert M. May. Princeton, N.J.: Princeton University Press, 1981.

———. *Origin of Species*. Edited with an introduction by J. W. Burrow. New York: Penguin Books, 1968.

Degler, Carl. *In Search of Human Nature: The Decline and Revival of Darwinism in American Social Thought*. New York: Oxford University Press, 1991.

Eldredge, N., and Steven Gould. "Punctuated Equilibrium: An Alternative to Phyletic Gradualism." In *Models in Paleobiology*, edited by J. S. Schopf, 82-115. San Francisco: Freeman & Cooper, 1972.

Elster, John. "The Possibility of Rational Politics." In *Political Theory Today*, edited by David Held, 139-40. Stanford, Calif.: Stanford University Press, 1991.

Gadamer, Hans. *Philosophical Hermeneutics*. Translated and edited by David E. Linge. Berkeley, Calif.: University of California Press, 1977.

———. *Truth and Method*. New York: Crossroad Press, 1982.

Galston, William. *Liberal Purposes: Goods, Virtues, and Diversity in the Liberal State*. New York: Cambridge University Press.

Gould, Steven. *Ever Since Darwin: Reflection in Natural History*. New York: W. W. Norton, 1977.

———. *Full House: The Spread of Excellence from Plato to Darwin*. New York: Three Rivers Press, 1996.

———. *The Mismeasure of Man*. New York: W. W. Norton, 1981.

———. *Urchin in the Storm: Essays about Books and Ideas*. New York: W. W. Norton, 1987.

Hallowell, John. *Main Current in Modern Political Thought*. New York: Henry Holt, 1950.

Hofstadter, Richard. *Social Darwinism in American Thought*. New York: George Braxiller, 1955.

Huxley, Julien. "Evolutionary Ethics." In *Darwin*, edited by Philip Appleman. New York: W. W. Norton, 1979.

Kolberg, Lawrence O. *Essays on Moral Development*. Vol. II. *The Psychology of Moral Development*. San Francisco: Harper & Row, 1984.

Lennox, James. "Aristotle on the Biological Roots of Virtue." In *Biology and the Foundation of Ethics*, edited by Jane Maienschen and Michael Ruse. New York: Cambridge University Press, 1999.

Lewontin, R. C. *Biology As Ideology*. New York: Harper & Row, 1992.

Lewontin, R. C., Steven Rose, and Leon Kamin. *Not in Our Genes: Biology, Ideology, and Human Nature*. New York: Pantheon Books, 1984.

Lumsden, Charles, and Edward O. Wilson. *Promethian Fire: Reflection on the Origin of Man*. Cambridge, Mass.: Harvard University Press, 1983.

Mayr, Ernst. "Biology, Pragmatism and Liberal Education." In *Education and Democracy: Reintegration of Liberal Learning in America*, edited by Robert Urrill. New York: College Board, 1997.

———. *This Is Biology: The Essence of the Living World*. Cambridge, Mass.: Harvard University Press, 1997.

———. *Toward a New Philosophy of Biology: Observations of an Evolutionist*. Cambridge, Mass.: Cambridge University Press, 1988.

McIntyre, Alasdair. *After Virtue: A Study in Moral Theory*. Notre Dame, Ind.: University of Notre Dame Press, 1981.

Midgley, Mary. *Beast of Man: The Roots of Human Nature*. New York: Routledge, 1979.

Nussbaum, Martha. *The Fragility of Goodness*. New York: Cambridge University Press, 1993.

Putnam, Hilary. "Aristotle after Wittgenstein." In *Words and Life: Hilary Putnam*. Cambridge, Mass.: Harvard University Press, 1994.

———. "The Dewey Lectures 1994." *Journal of Philosophy* (September 1994).

———, with Martha Nussbaum. "Changing Aristotle's Mind." In *Words and Life: Hilary Putnam*. Cambridge, Mass.: Harvard University Press, 1994.

———. *The Many Faces of Realism*. Lasalle, Ill.: Open Court, 1987.

————. *Realism, Truth and History.* New York: Cambridge University Press, 1989.

Rawls, John. *A Theory of Justice.* Cambridge, Mass.: Cambridge University Press, 1999.

————. *Political Liberalism.* New York: Columbia University Press, 1993.

Ruse, Michael. *Darwin Defended: A Guide to the Evolution of Controversies.* London: Addison-Wesley Publishing, 1982.

————. *The Darwinian Paradigm: Essay on Its History, Philosophy, and Religious Implications.* New York: Routledge, 1989.

Sabine, George. *History of Political Theory.* New York: Henry Holt, 1950.

Salkever, Stephen. *Finding the Mean: Theory and Practice in Aristotle's Political Philosophy.* Princeton, N.J.: Princeton University Press, 1990.

Searle, John. *Mind, Language and Society.* New York: Basic Books, 1998.

Taylor, Charles. *Philosophy and the Human Sciences, Philosophical Papers.* Vol. II. Cambridge: Cambridge University Press, 1985.

Waddington, C. W. *The Ethical Animal.* Chicago: University of Chicago Press, 1960.

Wallach, John. "Contemporary Aristotelianism." *Political Theory* (November 1992).

Wilson, E. O. *Consilience: The Unity of Knowledge.* New York: Alfred Knopf, 1998.

————. "Resuming the Enlightenment Quest." *Wilson Quarterly* (winter 1998).

————. *Human Nature.* Cambridge, Mass.: Harvard University Press, 1978.

————. "Human Decency As Animal." *New York Times Magazine.* October 19, 1975.

Yak, Bernard. *The Problem of a Political Animal.* Berkeley: University of California Press, 1993.

Index

About the Author

Terry Hoy is professor emeritus of political science at Simpson College, Indianola, Iowa. How is the author of several books in the area of political theory, including *The Political Philosophy of John Dewey, Towards a Naturalistic Political Theory: Aristotle, Hume, Dewey, Evolutionary Biology, and Deep Ecology.*